
STRUCTURES AND SOLID BODY MECHANICS DIVISION

General Editor: PROF. B. G. NEAL

Model Analysis of Plane Structures

Frontispiece

Model Analysis of Plane Structures

by

T. M. CHARLTON

PERGAMON PRESS

OXFORD · LONDON · EDINBURGH · NEW YORK
TORONTO · PARIS · FRANKFURT

Pergamon Press Ltd., Headington Hill Hall, Oxford
4 & 5 Fitzroy Square, London W.1

Pergamon Press (Scotland) Ltd., 2 & 3 Teviot Place, Edinburgh 1

Pergamon Press Inc., 44–01, 21st Street, Long Island City, New York 11101

Pergamon of Canada Ltd., 6 Adelaide Street East Toronto, Ontario

Pergamon Press S.A.R.L., 24, Rue des Écoles Paris 5^e

Pergamon Press GmbH, Kaiserstrasse 75, Frankfurt-am-Main

First edition 1966
Library of Congress Catalog Card No. 65-26346

Printed in Great Britain in the City of Oxford at The Alden Press

(2330/66)

CONTENTS

PREFACE

WITH the present-day preoccupation with automatic computation, there is a danger that insufficient attention will be given to devices of structural analysis which still have much to offer to the student and practising engineer. Such devices include the simple techniques of analysis plane structures with the aid of scale models with which this book is concerned. These techniques can be used successfully in the drawing office with the aid of ordinary household equipment such as a fretsaw for cutting models from celluloid sheet. It is noteworthy that model analysis is particularly useful for dealing with structures for which economy dictates members of non-uniform section. Moreover, it is easy to reduce the sections of the members of a model to investigate the possibility of saving of weight of metal or concrete.

Besides, elementary theory of similarity model analysis is based upon well-known principles of statics and linear elasticity. The technique whereby influence lines for forces and bending moments may be determined depends upon application of the principle of virtual work to linear systems. Apart from its value as an analytical device, this technique is valuable as a means of calling attention to features of the principle of virtual work which, it is believed, should be part of the mental equipment of students of civil and structural engineering.

Model analysis of the kind considered herein is a means of replacing the numerical work of formal structural analysis by mechanical operations. It utilizes the same assumptions as formal analysis and offers nothing more in the end-product, but the user gains unique insight into structural behaviour. Except for the purpose of necessary preliminary analysis, the techniques of model analysis described are not concerned with structural stability.

The author is indebted to the Publisher for obtaining the publication rights in respect of his earlier book on this subject, *Model*

Analysis of Structures (Spon, London, 1954), which is replaced by the present work. He is also indebted to the Director of the National Physical Laboratory for permission to include the illustration which forms the frontispiece and which provides an example of a laboratory model study.

Belfast, 1965 T. M. C.

CHAPTER 1

INTRODUCTION TO MODEL ANALYSIS

IT IS some forty years since accounts of model analysis of plane frameworks as an alternative to purely mathematical analysis were published by Professor G. E. Beggs of Princeton University in America, and others. At that time the analysis of statically indeterminate frameworks of the kind encountered in engineering practice was a forbidding task. Though the use of slope–deflection relationships for rigidly jointed frames had been systematized by Bendixen and treated in his book published in 1914,* distribution or relaxation methods were unknown and much research was devoted to alleviating the task of the structural analyst. Beggs's idea of putting Müller–Breslau's principle into use physically or experimentally for the purpose of obtaining influence lines for statically indeterminate quantities of linear plane frameworks by means of celluloid scale models received widespread acclaim. Unfortunately, however, the complexity and tediousness of the experimental technique soon put Beggs's apparatus into the teaching and research laboratory rather than the design office. Model analysis has never completely recovered from this initial setback in spite of the fact that it is well known that its more recent development has led to the elimination of complicated and tedious techniques and brought it within the scope of the design office. Moreover, additional approaches to model analysis, including the use of simple loading devices and the moment indicator, have

Die Methode der Alpha-gleichungen zur Berechnung von Rahmenkonstruktionen, Berlin, 1914.

widened its usefulness. There are numerous instances of frameworks of what might be called "intermediate" complexity for which model analysis is valuable in spite of the modern trend to automatic computation. (Model analysis is an alternative to mathematical analysis and not, as is sometimes supposed, only a means of checking analysis, though it can be useful in this way in some circumstances.) Use of scale models also provides unique experience and insight relating to the behaviour of structures generally and its value in this respect alone for teaching purposes, cannot be over-emphasized.

So-called indirect analysis of structures with linear characteristics by models has been developed from the technique proposed by Professor G. E. Beggs, being concerned with obtaining influence lines for forces and bending moments. There is no need for precision measuring equipment, and models of plane portal frameworks which fulfil the specified conditions for flexural similarity can be cut from sheet celluloid (or even cardboard) which have the requisite linear elastic properties. The whole of the work of making the models can, if necessary, be done in a design-office, since only very simple hand-tools are necessary. A wooden baseboard, such as a large drawing board, is all that is needed for the mounting of models for testing. The method is then mainly suitable for plane frames or structures which can be conveniently considered for analytical purposes as consisting of a number of plane frames connected together.

When a model has been fixed to a baseboard in such a way that the conditions specified for the supports of the prototype are reproduced, displacements are systematically introduced in turn at the supports or within members to obtain deformation curves which represent the influence lines for reactions and internal forces and bending moments, respectively, in accordance with Müller–Breslau's principle. Thus, forces and bending moments which cannot be found by the principles of statics alone can be obtained for any condition of loading of the prototype. The simplicity of the technique resides in the use of relatively large displacements and the absence of loading devices.

So-called direct analysis of structures by scale model tests, on the other hand, is concerned with obtaining strains or deformation due to specified loading, including influence lines for deformation. For plane frameworks a simple direct model testing technique is available for which the same kind of model and mounting arrangements may be used as for the indirect technique, in the manner described in Chapter 6. Loading of plane celluloid models may be achieved by cantilever type balances made of the same material and bending moments in members may be obtained by the "moment indicator" (Chapter 6). Again, precision devices for measuring deformations are unnecessary; pencil and paper and a graduated scale are sufficient, generally. For some types of framework, such as the Vierendeel bridge girder, direct model analysis of the kind described is superior to indirect analysis.

Direct model analysis includes as well, of course, the kind of model study which is sometimes desirable for investigating the strength and economy of complex structural forms such as shells, domes and concrete highway bridges as well as the effects of wind on tall buildings. Models of these structures might be made of perspex, micro-concrete or possibly, expanded polystyrene for testing under direct loading to determine strains and deflections. A perspex model of a reinforced concrete tower building is shown undergoing test for deformation due to wind loading in the Frontispiece (see Ref. 35 of Appendix I).

Early Approaches to Systematic Model Analysis

Three model-testing techniques which employ the indirect principle but which have never been widely accepted by practising engineers because of their general inconvenience were developed in the 1920's. The names of Beggs, Gottschalk and Reickhof are associated with them, respectively. Teaching institutions in many parts of the world have adopted one or other of these methods for purposes of instruction in structural behaviour.

Beggs's Apparatus

Professor G. E. Beggs seems to have been the first to introduce (1922) a systematic model technique of the indirect type for the analysis of linear structures. Beggs advocated the use of celluloid or cardboard models, for manipulation by a set of instruments to produce influence lines for forces and bending moments. These instruments included means of clamping models at supports in such a way that predetermined displacements could be introduced as desired, by changing certain elements of the clamps. A different type of clamp was used for connecting the adjacent ends of cut members to enable internal rotational, axial and shear displacements to be applied. The cutting of members after the application of the clamping device could be practised, if desired.

Very small displacements were used, their values being determined by the dimensions of the displacement elements which were inserted in the clamps. A microscope was used for measuring the deformations (i.e. influence line ordinates) of the model due to the displacements introduced. Sets of standard clamps and a microscope were made commercially as Beggs's apparatus, and the user had to make only the model to his own requirements from cardboard or celluloid. Fundamentally, the Beggs technique differs from that of the indirect method described herein in respect of the magnitudes, method of introducing and measuring displacements and deformations. Beggs mounted his models horizontally and supported them at a number of points on ball-bearing mountings.

Gottschalk's Apparatus—"The Continostat"

Otto Gottschalk of Buenos Aires published his version of the indirect model method in 1926. His apparatus, which he called the "Continostat", consisted of a variety of steel strips or splines, together with means of clamping them together and mounting them to represent portal structures and continuous beams. For the purpose of obtaining influence lines, arbitrarily chosen large displacements could be introduced both internally and externally to the models, which were mounted vertically on a baseboard covered with paper to enable the deformed shapes or influence lines to be recorded. By virtue of the magnitudes of the displacements introduced the ordinates of the influence lines so obtained could be measured by ordinary scales with the naked eye.

The Continostat was not completely adaptable because of the inadequacy of the standard sets of splines for the representation of some structural members. Various second moments of area could be reproduced by suitable combinations of splines, but clearly the number of such combinations is limited.

Gottschalk's Continostat was made commercially and seems to have been one, if not the earliest, model apparatus for use in conjunction with so-called large displacements.

Rieckhof's Apparatus—"The Nupubest"

Rieckhof's apparatus, which he called the "Nupubest" had much in common with Gottschalk's Continostat. It consisted of a set of steel splines with clamping devices for the construction of models, together with means of supporting them and applying internal and external displacements. In addition to producing influence lines by the indirect method using large displacements, Rieckhof proposed an alternative approach for the analysis of structures made up of straight members. This involved the determination of points of contraflexure by the use of his apparatus and completing the analysis by means of the well-known relationships for beams in bending.

Rieckhof used his Nupubest in the horizontal plane, and supported it on ball-bearing mountings to eliminate friction. The limitations of the apparatus are similar to those of Gottschalk's apparatus.

Limitations and Accuracy of Model Analysis

Simple model analysis of plane frameworks usually suffers from the same kind of idealization as mathematical analysis. For example, it is difficult to reproduce the load–deflection characteristics of joints and foundations and it is usual to assume full rigidity or freedom as for mathematical analysis. In other words, model analysis of the kind with which this book is concerned is precisely equivalent to mathematical methods both in respect of assumptions and objective.

The accuracy of model analysis based upon the same assumptions as the corresponding mathematical analysis, is sufficiently good for practical purposes, being generally within 5 per cent or at the most 10 per cent, of the correct values obtained by mathematical analysis. Much depends, of course, upon the care with which models are devised and used, though no particular manual skill is required to achieve the degree of accuracy described.

CHAPTER 2

SIMILARITY OF STRUCTURES

Two geometrically similar structures are flexurally* similar if individual deflections, which are related by the geometrical scale factor between the structures both in magnitude and in position, are associated with deformed shapes of the structures which are also to scale, i.e. geometrically similar. Thus similarity implies that the relevant quantities of the two systems can be related by fixed-scale factors. As the stresses in structures depend upon their deformations, flexural similarity implies that the stress distributions are similar for geometrically similar deformations. One of the earliest treatments of similarity of structures is due to Professor James Thomson, brother of Lord Kelvin, and Professor of Civil Engineering in the Queen's College of Belfast from 1857 until 1873 (Appendix I, Ref. 1).

The approach to flexural similarity given below consists essentially of determining directly the conditions for the properties of deformation of two geometrically similar structures to be similar. The approach which is afforded by the methods of dimensional analysis, however, verifies rigorously that geometrical similarity is a prior requirement for the flexural similarity of structures. This is illustrated with reference to the simple example of a single beam.

General Expression for the Deflection of a Structure

The deflection of a linearly elastic structure at any point i, due

* The term "flexural" as used in this book relates to deformations of structures or their members generally, due to load: for example, the deformation of a truss due to axial strain of its members or the deformation in bending of a beam.

to a force F_j applied at any point j, can be expressed generally as follows:

$$\Delta_i = F_j\left(K_1 \frac{l}{EA} + K_2 \frac{Al}{GI} + K_3 \frac{l^3}{EI} + K_4 \frac{l^3}{GQ}\right), \tag{2.1}$$

by referring to one of its members of length l, cross-sectional area A, second moment of area I, torsion constant Q, Young's modulus E and modulus of rigidity G, and using appropriate constants K_1, K_2, K_3 and K_4 which take into account the layout geometry of the structure, the position of the load F_j together with its effect within the structure and the contributions of the other members due to their axial deflection, shear deflection, bending and twisting respectively. The quantity F_jl/EA represents deflection due to direct extension or compression; F_jAl/GI represents deflection due to shear; F_jl^3/GQ represents deflection due to torsion and F_jl^3/EI represents deflection due to bending.

By eqn. (2.1) the deflection of the structure at any point i when unit load is applied at the point j, i.e. the flexibility at i with respect to j, is therefore

$$a_{ij} = \frac{\Delta_i}{F_j} = K_1 \frac{l}{EA} + K_2 \frac{Al}{GI} + K_3 \frac{l^3}{EI} + K_4 \frac{l^3}{GQ}. \tag{2.2}$$

Similar expressions can be written for the other flexibility coefficients of the structure, namely, $a_{11}, a_{12}, a_{13}, \ldots; a_{21}, a_{22}, a_{23}, \ldots$, etc., where in general $a_{ij} = a_{ji}$ by the reciprocal theorem.

Conditions for Flexural Similarity

For two geometrically similar structures, for example, model and prototype, to be flexurally similar, the flexibilities a which describe the respective deflection characteristics must bear the same ratio one to another for both structures. That is

$$\frac{a_{11}}{a_{12}} = \frac{a'_{11}}{a'_{12}}; \frac{a_{21}}{a_{22}} = \frac{a'_{21}}{a'_{22}}; \ldots; \frac{a_{i1}}{a_{ij}} = \frac{a'_{i1}}{a'_{ij}}; \text{ etc.,} \tag{2.3}$$

from which it follows that

$$\frac{a_{11}}{a'_{11}} = \frac{a_{12}}{a'_{12}} = \frac{a_{22}}{a'_{22}} = \ldots = \frac{a_{ij}}{a'_{ij}}, \tag{2.4}$$

where the prime denotes the quantities relating to the second structure.

For the similarity conditions of eqn. (2.3) to be fulfilled in general, it follows from eqn. (2.2) that, besides the dimensionless quantities (constants) K_1, K_2, K_3 and K_4 being the same for the structures, the dimensionless quantity E/G must be the same for both, i.e.

$$\frac{E}{G} = \frac{E'}{G'}. \tag{2.5}$$

This means that Poisson's ratio ($\nu = E/2G - 1$) must be the same for the materials of the prototype and its model respectively, unless shear and torsion are negligible, when E is the only significant elastic constant and may be different for the two materials. (It is usually possible to neglect the effect of shear on the deflection of plane frames).

It should be noted from eqn. (2.2) that if the deflection due to axial, torsion and shear effects (as distinct from that due to bending) can be neglected, as for plane portal frames, the relevant second moment of area I is the significant geometrical property of the cross-section of members in respect of flexibility or stiffness. Again, if bending, torsion and shear are negligible as for pin-jointed structures, A is the only cross-sectional property of the members which need appear in the analysis. These considerations are relevant when designing a model, since they allow some freedom in the selection of the cross-sectional forms of members.

Theory of Dimensions and Similarity

Any equation which describes a physical phenomenon must be dimensionally correct; that is, the dimensions (e.g. length L, mass M, time T) of both sides of the equation must be the same.

By Newton's second law, force is equal to mass multiplied by acceleration, the dimensions of the latter being L/T^2. If, therefore, an equation specifies equilibrium, both sides of the equation must have the dimensions of force $F = ML/T^2$. The use of dimensional analysis in this manner for checking equations of physical phenomena is well known and of great value.

The dimensional approach can be extended to provide a means of determining the manner in which the variables influence physical phenomena. A valuable device in this respect is Buckingham's "pi" theorem.

Buckingham's theorem states that, if in the equation representing a physical phenomenon, any variable X_1, depends solely upon *independent* variables $X_2, X_3, \ldots, X_n$, such that

$$X_1 = f(X_2, X_3, X_4, \ldots, X_n), \tag{2.6}$$

or

$$\phi(X_1, X_2, X_3, \ldots, X_n) = 0, \tag{2.7}$$

then, if μ is the *minimum* number of dimensions involved* eqn. (2.7) can be expressed alternatively in terms of $(n - \mu)$ dimensionless quantities. Thus:

$$\psi(\pi_1, \pi_2, \ldots, \pi_{n-\mu}) = 0, \tag{2.8}$$

where π is a dimensionless quantity or a ratio which is a function of $(\mu + 1)$ of the variables. It follows that in general there must be μ unknown exponents of the variables in each π, but since the number of fundamental dimensions is μ, each exponent can be found dimensionally, that is, without knowledge of the functions ϕ or ψ (see, for example, eqns. (2.12) and (2.13)). The pi theorem thus indicates how the phenomenon under consideration depends upon each variable individually. It provides no information concerning the functional relationship between the variables; other means, such as experiment are necessary for this purpose. The value of the theorem is dependent upon all of the significant variables being taken into account and use of the minimum possible number of dimensions. In determining the dimensions

* See Appendix I, Ref. 8, pp. 171–81.

and significant variables it is important to combine any which always occur together. For example, in statical problems, the dimension of time occurs only in the dimensional representation of force $F = (ML/T^2)$ so that the minimum number of dimensions is two, i.e. F and L. Also, in the simple theory of flexure of beams where E and I always occur together, EI should be treated as a single variable B otherwise misleading results might be obtained in applying the pi theorem.

The pi theorem further provides a basis for the comparison of separate manifestations of a given phenomenon without reference to its governing equations, because the form of the π's is the same for each. Thus, when the behaviour of two systems is similar, the numerical values of their respective π's are the same. This is the basis of the general study of physical similarity.

Although the number of π's which is necessary to describe a given phenomenon is determined, each π can take several alternative forms depending upon the scheme which is adopted for their formulation. Again, if necessary, any one π can be replaced by itself combined in any way with one or more of the other π's. Thus, if π_1, π_2, π_3 and π_4 are applicable to a certain phenomenon, π_1 can be replaced by, say, $\pi_1' = \pi_1\pi_3/\pi_4$, if necessary, so that the set becomes π_1', π_2, π_3 and π_4.

Dimensional Analysis of Beam Flexure: Flexural Similarity

In order to illustrate the use of the pi theorem for structures, it is convenient to consider the deflection of a uniform beam (whose elasticity is linear) due to, say, a central concentrated load F. The independent variables by the simple theory of bending are the length of the beam l; its flexural rigidity $EI = B$; the load F and the central deflection Δ which it produces. The deflection at the centre of the beam can therefore be expressed as follows:

$$\Delta = \phi_1(l, B, F). \tag{2.9}$$

Eqn. (2.9) can be written alternatively as:

$$\Phi_1(l, B, F, \Delta) = 0. \tag{2.10}$$

Now the fundamental dimensions are M, L and T, but, as explained above, since in statical problems time never appears other than in the dimensional expression for force, it is legitimate to replace M and T completely by the dimension of force W treated as a fundamental quantity. Thus the minimum number of dimensions is two, so that the number of π's is two because there are four independent variables. Eqn. (2.10) can then be rewritten by application of Buckingham's theorem as follows:

$$\psi_1(\pi_1, \pi_2) = 0. \tag{2.11}$$

One way of finding two π's for the beam is to select two of the variables, e.g. B and l, and make them common to both of the π's, as follows:

$$\begin{aligned} \pi_1 &= l^{\alpha_1} \times B^{\beta_1} \times \Delta \\ \pi_2 &= l^{\alpha_2} \times B^{\beta_2} \times F. \end{aligned} \tag{2.12}$$

Now, Δ and l have the dimensions of length L, while B has the dimensions of $W \times L^2$, W being a fundamental dimension in this instance. Equations (2.12) can, therefore, be written in dimensional form as follows:

$$\begin{aligned} \pi_1 : 1 &= L^{\alpha_1} \times W^{\beta_1} \times L^{2\beta_1} \times L \\ \pi_2 : 1 &= L^{\alpha_2} \times W^{\beta_2} \times L^{2\beta_2} \times W. \end{aligned} \tag{2.13}$$

The figure one on the left-hand side of these equations signifies that the π's are dimensionless, so that the exponents α and β can be determined by the following simultaneous equations:

for L:

$$\begin{aligned} \alpha_1 + 2\beta_1 + 1 &= 0 \\ \alpha_2 + 2\beta_2 &= 0 \end{aligned} \tag{2.14}$$

for W:

$$\begin{aligned} \beta_1 &= 0 \\ \beta_2 + 1 &= 0 \end{aligned} \tag{2.15}$$

whence $\alpha_1 = -1, \beta_2 = -1$ and $\alpha_2 = 2$.

Therefore:

$$\begin{aligned} \pi_1 &= \frac{\Delta}{l} \\ \pi_2 &= \frac{Fl^2}{B}, \end{aligned} \tag{2.16}$$

and eqn. (2.11) can be replaced by the equation:

$$\psi_1\left(\frac{\Delta}{l}, \frac{Fl^2}{B}\right) = 0, \tag{2.17}$$

or

$$\Delta = l\psi_2\left(\frac{Fl^2}{B}\right). \tag{2.18}$$

It should be noted that when two variables of the same dimension, e.g. Δ and l occur in an expression for a π, it can be written by inspection as the ratio of the variables, i.e. Δ/l.

While the above procedure for the beam is compatible with the elementary Bernoulli–Euler bending theory, it is not exact in that shear and lateral effects are ignored completely by omitting the modulus of rigidity G or Poisson's ratio ν and using B instead of E and cross-sectional dimensions, e.g. t and d. In order to take these additional effects into account it is necessary then to introduce the modulus of rigidity and replace B by E, t and d. The new set of π's can be written by inspection with one exception (π_4) as follows:

$$\pi_1 = \frac{t}{l}; \pi_2 = \frac{d}{l}; \pi_3 = \frac{\Delta}{l}; \pi_4 = \frac{F}{Ed^2}; \pi_5 = \frac{G}{E}. \tag{2.19}$$

π_4 can be verified by a procedure similar to that adopted above. In accordance with the pi theorem, the new variables, E, G, t and d increase the number of π's by three because now $n = 7$ while the minimum number of basic dimensions is still two. The deflection equation can be written:

$$\Delta = l\psi\left(\frac{t}{l}, \frac{d}{l}, \frac{F}{Ed^2}, \frac{G}{E}\right). \tag{2.20}$$

The general conditions for the flexural similarity of two beams are that the π's of eqn. (2.19) shall have the same numerical value for both. It is apparent, therefore, that the beams must be geometrically similar as dictated by $\pi_1 = t/l$ and $\pi_2 = d/l$. Furthermore, their respective materials of construction must obey Hooke's law, and, in general, have the same value of Poisson's ratio as specified by

$$\pi_5 = \frac{G}{E}.$$

The remaining π's, namely, $\pi_3 = \Delta/l$ and $\pi_4 = F/Ed^2$, provide a means of relating the deflections and loads of the two beams respectively. Thus, if the ratio Δ/l, is the same for the two beams, the loads can be related by π_4 for similarity, i.e.

$$\frac{F}{Ed^2} = \frac{F'}{E'd'^2},$$

whence

$$F = \left(\frac{d}{d'}\right)^2 \frac{E}{E'} F'. \tag{2.21}$$

If only simple bending is significant, however, eqns. (2.16) are applicable and if $\Delta/l = \Delta'/l'$:

$$F = \left(\frac{l'}{l}\right)^2 \frac{B}{B'} F'. \tag{2.22}$$

The more general beam problem serves to confirm the requirements for the flexural similarity of linear structures, namely, similar geometry and equal values of Poisson's ratio. Sometimes the weight of structural members is important because of its influence upon deformation and stability generally. This factor may be taken into account by including the variables which represent the density of the material of the members and the acceleration due to gravity, respectively, in the dimensional analysis.

Relationships Between Model and Prototype Quantities

When the model is $1/s$ times as large as the prototype the following equations for relating the model quantities to those of the prototype are obtained, provided that the prototype and model deflections satisfy the requirements of similarity by being in the ratio of the layout scale factor s, i.e.

$$\Delta = s\Delta', \tag{2.23}$$

where the prime refers to the model. Thus, a deflection Δ' of the model is equivalent to a deflection $\Delta = s\Delta'$ of the prototype at the corresponding point.

By eqns. (2.1), (2.21) or (2.22):

$$F = s^2 \frac{E}{E'} F', \tag{2.24}$$

which is the relationship between the forces applied to the prototype and its model when the respective deflections produced are in accordance with eqn. (2.23). On the same basis it follows that for stresses:

$$\sigma = \frac{E}{E'} \sigma', \tag{2.25}$$

(i.e. identical strains) by making use of eqn. (2.24) and the fact that stress is equal to force divided by area. Thus, if the model is made of the same material as the prototype the stresses are the same for both when the deflections are in accordance with eqn. (2.23).

By eqn. (2.2) or eqns. (2.23) and (2.24) the flexibilities are related as follows:

$$a = \frac{1}{s} \frac{E'}{E} a'. \tag{2.26}$$

It should be noted that, although the ratio of the moduli of elasticity appears in eqns. (2.24) to (2.26) inclusive, where shear and torsion are involved the ratio G'/G could be used equally well, because then one of the conditions for flexural similarity is that $E/G = E'/G'$.

If the model is constructed with layout dimensions $1/s$ times as large as those of the prototype, while the cross-sectional dimensions of its members are $1/r$ times those of the prototype, on the basis of only bending or torsion or both being significant, the scaling formulae become:

$$\Delta = s\Delta', \tag{2.23}$$

$$F = \frac{r^4}{s^2}\frac{E}{E'}F', \tag{2.27}$$

$$\sigma = \frac{r^2}{s^2}\frac{E}{E'}\sigma', \tag{2.28}$$

$$a = \frac{s^3}{r^4}\frac{E'}{E}a'. \tag{2.29}$$

These formulae can be verified by referring to eqn. (2.1) which reduces to the following form when axial and shear effects are negligible:

$$\Delta = F\left(K_3\frac{l^3}{EI} + K_4\frac{l^3}{GQ}\right), \tag{2.30}$$

$$\Delta' = F'\left(K_3\frac{l'^3}{E'I'} + K_4\frac{l'^3}{G'Q'}\right). \tag{2.31}$$

By putting $\Delta = s\Delta'$ and remembering that $a = \Delta/F$; $l = sl'$; $I = r^4I'$ and $Q = r^4Q'$ eqns. (2.27) to (2.29) inclusive are obtained.

If, on the other hand, only deflection due to the axial forces in members is significant, the scaling formulae for $\Delta = s\Delta'$ are:

$$F = r^2\frac{E}{E'}F', \tag{2.32}$$

$$\sigma = \frac{E}{E'}\sigma', \tag{2.33}$$

$$a = \frac{s}{r^2}\frac{E'}{E}a', \tag{2.34}$$

because reference to eqn. (2.1) as above, shows that in this instance:

$$\Delta = K_1 \frac{Fl}{EA}, \tag{2.35}$$

and

$$\Delta' = K_1 \frac{F'l'}{E'A'}, \tag{2.36}$$

where

$$l = sl'; \; A = r^2 A'.$$

It should again be noted that the requirement of $E/G = E'/G'$ (equal values of Poisson's ratio) is unimportant when only bending or axial effects of members are significant, because G does not appear in the analysis. It is also important to note that eqn. (2.23) is the basis of the force, and stress scaling formulae.

CHAPTER 3

THEORY OF MODEL ANALYSIS

METHODS of analysing frameworks with linear characteristics by means of scale models are essentially two-fold. The so-called direct method enables deformations caused by specified loading to be obtained directly. It is also a means of obtaining influence lines for deflection. The so-called indirect method is concerned with obtaining influence lines for forces (and bending moments) at supports or in members of a framework. The direct method embraces techniques based on slope–deflection relationships for members in bending for determining bending moments in members of rigidly jointed frameworks. Otherwise it utilizes the reciprocal theorem due to Maxwell and Betti when influence lines for deflection of frameworks are required. The indirect method is based on Müller–Breslau's principle and consists essentially of physical application of the principle to scale models. The theory of indirect model analysis is considered first in this chapter.

Influence Lines for Forces: Müller–Breslau's Principle

Müller–Breslau's principle is applicable to statically determinate structures as well as to statically indeterminate structures whose load-deflection characteristics are linear. That is, it is applicable to structures for which the concept of influence lines is valid by virtue of the principle of superposition of forces. It states that the influence line for the force (or moment) in any member or support of a structure is represented by the displaced or deformed configuration of the structure due to the application, within the member or at the support, of a *small* arbitrary displacement in the line of action of the force. Müller–Breslau's principle can be verified by means of the principle of virtual work but an alternative derivation using the reciprocal theorem of Maxwell and Betti is possible for statically indeterminate systems within the range of linear elasticity.

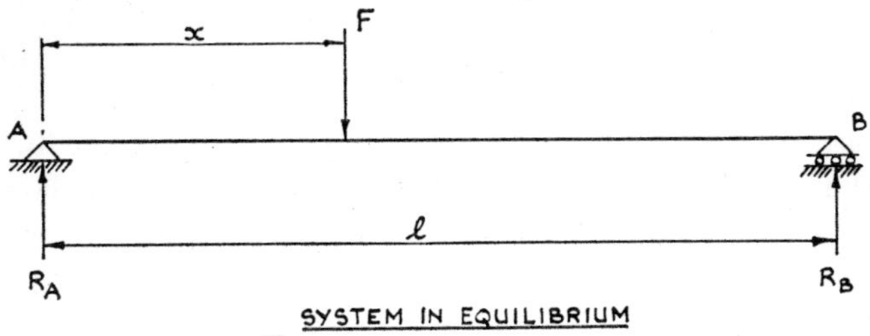

FIG. 1 (*a*)

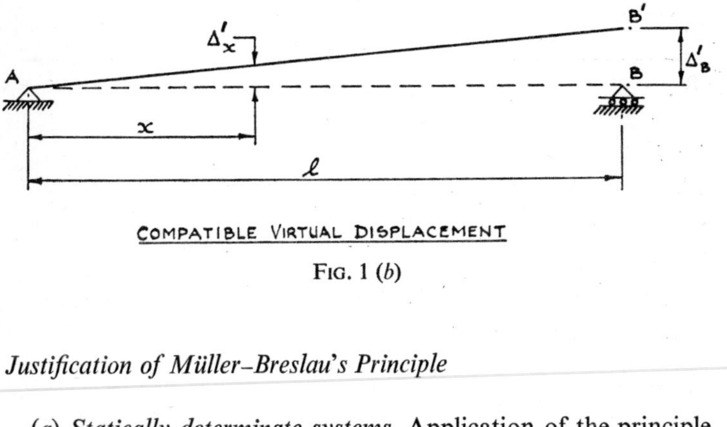

FIG. 1 (*b*)

Justification of Müller–Breslau's Principle

(*a*) *Statically determinate systems.* Application of the principle of virtual work (Appendix II) to the force system of a simply supported beam loaded as shown in Fig. 1(*a*) for the small imaginary or virtual displacement shown in Fig. 1(*b*) gives:

$$\underbrace{\overbrace{R_A O + R_B \Delta'_B - F\Delta'_X}^{\text{System in equilibrium}}}_{\text{Compatible displacements}} = 0, \tag{3.1}$$

whence

$$R_B = \frac{\Delta'_X}{\Delta'_B} F, \tag{3.2}$$

or since $\Delta'_X = \Delta'_B x/l$

$$R_B = \frac{x}{l} F. \tag{3.3}$$

Thus, the displaced form AB' represents the influence line for R_B so long as Δ'_B is much smaller than l, that is, so long as the dis-

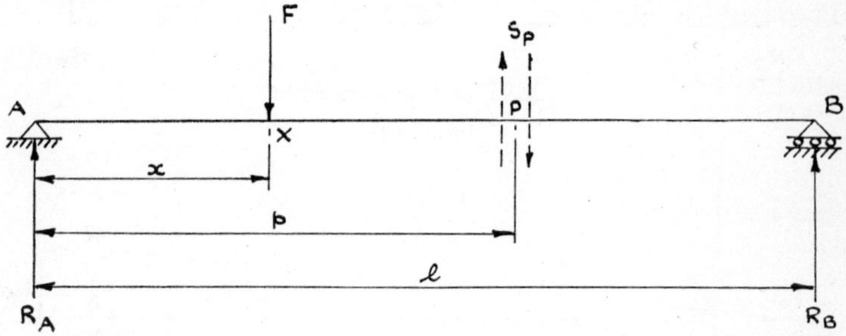

FIG. 2 (*a*)

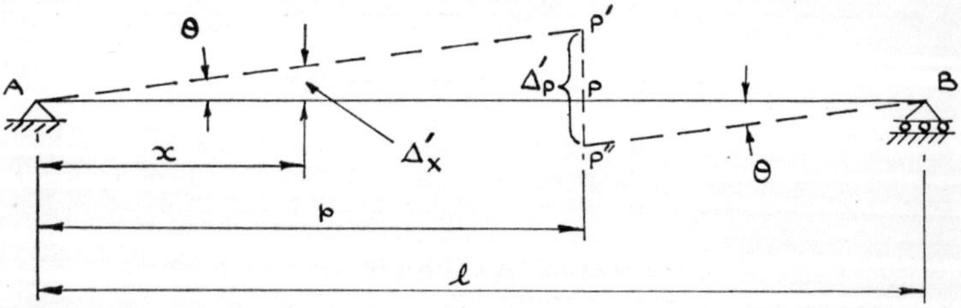

FIG. 2 (*b*)

placement is compatible with the geometry of the beam system as loaded. This is in accordance with Müller–Breslau's principle.

The influence line for the shear force at some point P of a simply supported beam (Fig. 2) may be obtained by considering a small shear displacement Δ'_P to be introduced at P, as shown in Fig. 2(*b*). (When considering the effect of the shear displacement it is helpful to think of the beam "cranked" at P by a small amount of Δ'_P and then placed upon its supports.) The resulting shape $AP'P''B$ represents the influence line for shear force at P. The shear force S_P at point P as shown in Fig. 2(*a*) is related to a concentrated load F applied at any point X by applying the principle of virtual work to these forces in equilibrium with the virtual displacements shown in Fig. 2(*b*), thus:

$$-S_P\Delta'_P - F\Delta'_X = 0, \tag{3.4}$$

if positive shearing force S_P is as shown in Fig. 2(*a*).

When $x \leqslant p$, $\Delta'_X = (x/l)\Delta'_P$ therefore:

$$S_P = -\frac{x}{l}F, \tag{3.5}$$

and when $x > p$, $\Delta'_X = -(l-x)\Delta'_P/l$, so that:

$$S_P = \frac{(l-x)}{l}F, \tag{3.6}$$

which may be verified by elementary statics since $R_B = (x/l)F$.

The influence line for the bending moment at P of the simply supported beam (Fig. 3) may be obtained similarly by considering a small relative rotation of the two parts of the beam AP and PB at P, as shown in Fig. 3(*b*). Thus, the deformed shape $AP'B$ represents the influence line for bending moment at P, since, by the principle of virtual work the sagging bending moment M_P at P due to a concentrated load F at any point X, as shown in Fig. 3(*a*), is:

$$M_P = \frac{\Delta'_X}{\Delta'_\theta}F, \tag{3.7}$$

where Δ'_θ is the virtual relative rotation at P causing the beam to hog, as shown in Fig. 3(*b*), and Δ'_X is the resulting displacement of

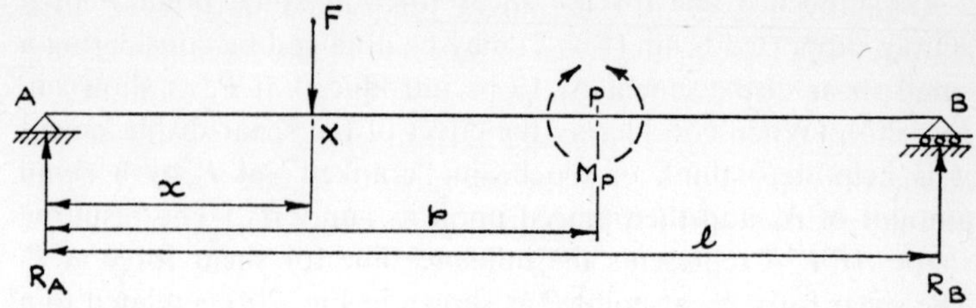

FIG. 3 (*a*)

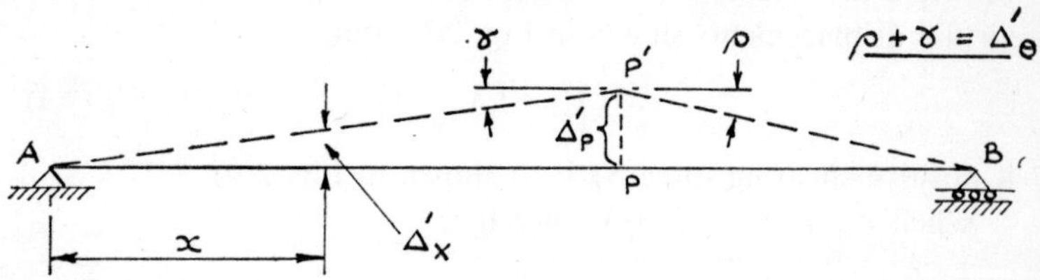

FIG. 3 (*b*)

point X. Δ'_θ may be expressed in terms of the resulting vertical displacement Δ'_P of P as follows:

$$\Delta'_\theta = \frac{\Delta'_P}{p} + \frac{\Delta'_P}{l-p}, \tag{3.8}$$

where $p = AP$. Therefore:

$$\Delta'_\theta = \frac{l}{p(l-p)}\,\Delta'_P, \tag{3.9}$$

so that by substituting in eqn. (3.7) and noting that $\Delta'_X/\Delta'_P = x/p$ for $x \leqslant p$:

$$M_P = \frac{x(l-p)}{l}\,F, \tag{3.10}$$

while if $x > p$, $\Delta'_X/\Delta'_P = (l-x)/(l-p)$ and:

$$M_P = \frac{p(l-x)}{l}\,F, \tag{3.11}$$

as may be verified by elementary statics.

Again, for a statically determinate pin-jointed plane framework the influence line for the force in any one of the members can be obtained by considering the appropriate system of compatible small virtual displacements for the framework. For example, to obtain the influence line for the tension T_{PQ} in member PQ of the framework shown in Fig. 4(*a*) the compatible virtual displacement

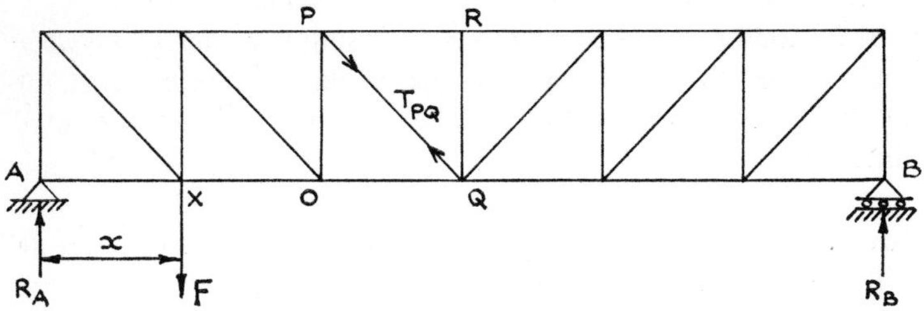

FIG. 4 (*a*)

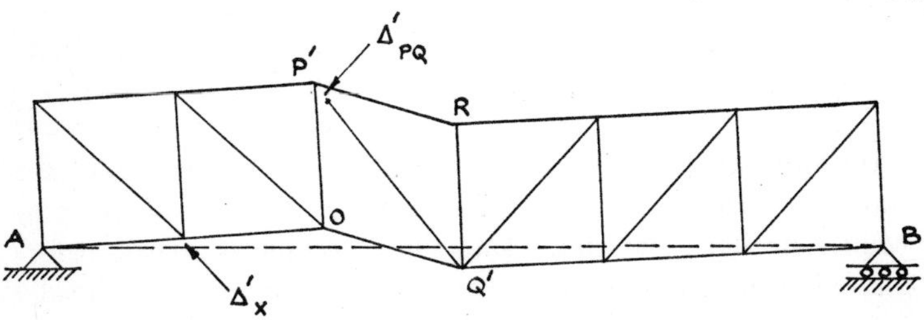

FIG. 4 (*b*)

system shown exaggerated in Fig. 4(*b*) is appropriate. A small virtual displacement Δ'_{PQ} is introduced to extend the line PQ causing (compatible) virtual displacement of the framework as shown. By the principal of virtual work, then, having regard to the fact that for the compatible virtual displacement chosen, virtual work is done on both F and T_{PQ}:

$$-F\Delta'_X - T_{PQ}\Delta'_{PQ} = 0, \tag{3.12}$$

whence

$$T_{PQ} = -\frac{\Delta'_X}{\Delta'_{PQ}} F. \tag{3.13}$$

As tension was assumed initially, the minus sign denotes that the member is in compression when the load is in the position shown.

The line $AOQ'B$ in relation to the baseline AB represents the influence line for the tension in member PQ for loading of the lower boom of the framework. Moreover, the deformed shape of the framework as a whole shown in Fig. 4(*b*) represents the complete influence line system for the force in member PQ.

If the load F acts between joints Δ'_X is simply measured between the baseline AB and the line of the displaced framework at the point of application of the load. Actually the load is distributed linearly between the joints on either side of the load, in this instance, by the connecting member acting as a simply supported beam. If this practical feature is taken into account in using an influence line it will be found that the axial force in the relevant member of the truss is the same as though the load is considered to act wholly at its actual point of application on the structure. (In order to avoid primary bending effects in members of trusses, secondary structure is usually provided to transfer loads to joints in the manner described, e.g. the deck in the case of a bridge.)

(*b*) *Statically indeterminate systems.* Whereas influence lines for statically determinate structures can be obtained directly by Müller–Breslau's principle using simple geometry, elastic deformation is involved when the principle is applied to statically indeterminate structures. Unless, therefore, scale models are used to obtain the required deformation curves or lines, tedious calculations may be necessary.

In order to demonstrate the validity of Müller–Breslau's principle for linear statically indeterminate frameworks, suppose the influence line for the member PQ of the plane statically indeterminate framework with linear elasticity shown in Fig. 5(*a*) is required. Introduction of a small extension in the line PQ will cause displacement throughout the framework as shown exaggerated in Fig. 5(*b*). By Müller–Breslau's principle this displaced form represents the required influence line when it is related to the original configuration of the framework. Its appearance is similar to that for the member PQ of the statically determinate system shown in Fig. 4(*b*), but now owing to the effect of the redundant members the ordinates of the two influence lines are not identical. In fact, introduction of a small extension in the line PQ of the statically indeterminate framework requires application of an external force F'_{PQ} between P and Q causing a force $-T'_{PQ}$ in

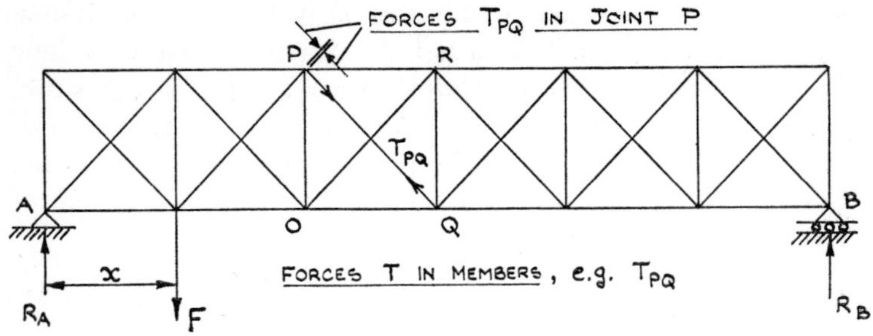

FIG. 5 (*a*)

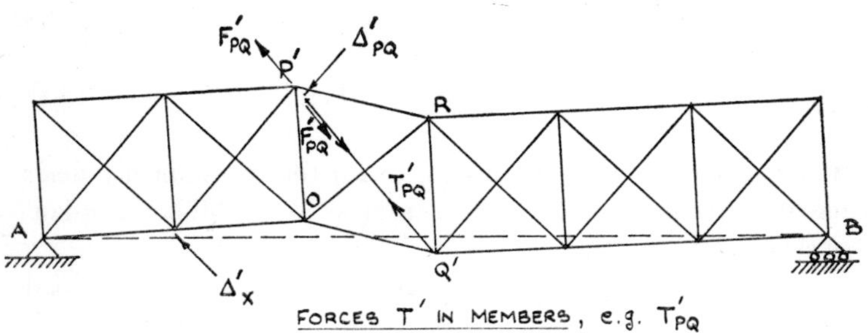

FIG. 5 (*b*)

member PQ and forces T' in the other members of the structure with corresponding changes in length e' of the members including PQ, as shown in Fig. 5(*b*). (The introduction of a displacement within a statically determinate system requires no force for the purpose.)

Applying the principle of virtual work to the system of forces in equilibrium shown in Fig. 5(*a*) for small virtual displacements equal to the displacements caused by the system of forces shown in Fig. 5(*b*) gives:

$$-F\Delta'_X - T_{PQ}\Delta'_{PQ} = \sum Te' = \sum T\frac{T'l}{AE}, \tag{3.14}$$

since $e' = T'l/AE$ for linear elasticity (l, A, E are length, cross-sectional area and Young's modulus of a member respectively) and where the summation embraces *all* members of the framework (including PQ). The term $\sum Te'$ denotes virtual work done on the members of the framework when T and e' are positive. Again, applying the principle of virtual work to the system of forces in equilibrium shown in Fig. 5(*b*) for small virtual displacements equal to the displacements caused by the system of forces shown in Fig. 5(*a*) gives:

$$O\Delta'_X = \sum T'e = \sum T'\frac{Tl}{AE}, \tag{3.15}$$

whence, by eqn. (3.14):

$$-F\Delta'_X - T_{PQ}\Delta'_{PQ} = 0, \tag{3.16}$$

or

$$T_{PQ} = -\frac{\Delta'_X}{\Delta'_{PQ}}F, \tag{3.17}$$

which is similar in form to eqn. (3.13) for the statically determinate system and is in accordance with Müller–Breslau's principle.

For statically indeterminate systems whose members resist loads by bending, corresponding relationships may be obtained. Thus, for the continuous beam shown in Fig. 6(*a*), introduction of a small displacement at support Q as shown in Fig. 6(*b*) gives a deformed configuration which represents the influence line for the force on that support. Verification may be obtained by applying the principle of virtual work, first to the force system of Fig. 6(*a*) using virtual displacements equal to the displacements caused by the force system shown in Fig. 6(*b*), then to the force system of Fig. 6(*b*) with virtual displacements equal to the displacements caused by the force shown in Fig. 6(*a*), as follows:

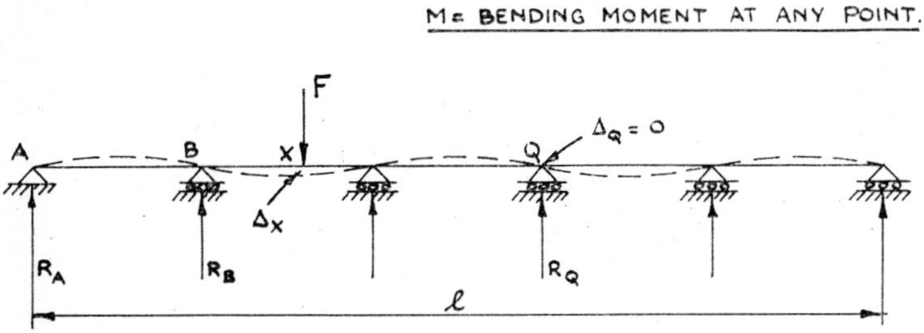

FIG. 6 (*a*)

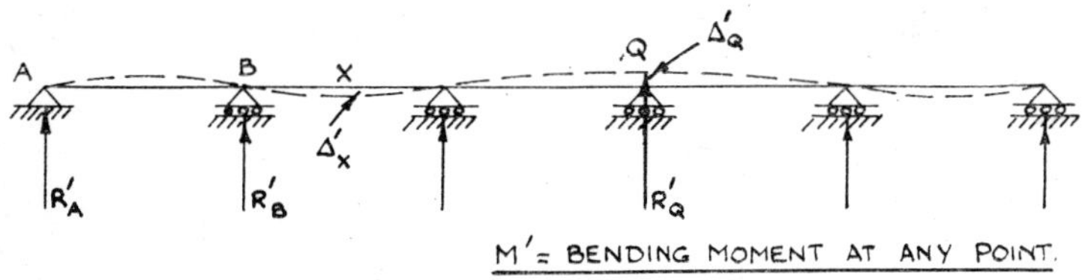

FIG. 6 (*b*)

System in equilibrium

$$F\Delta'_X + R_Q\Delta'_Q - \int_0^l M\,\frac{M'\,\mathrm{d}x}{EI} = 0, \tag{3.18*}$$

Compatible displacements

(since $M'\delta x/EI$ is the angle of bending of an element of beam δx long and of flexural rigidity EI, due to bending moment M') and:

$$O\Delta_X + R'_Q O - \int_0^l M'\,\frac{M\,\mathrm{d}x}{EI} = 0, \tag{3.19*}$$

* The terms $\int_0^l M\,\frac{M'\,\mathrm{d}x}{EI}$ are negative denoting virtual work done on the beam for positive values of M and M'.

respectively, whence: $$R_Q = -\frac{\Delta'_X}{\Delta'_Q} F, \tag{3.20}$$

by eqn. (3.18) after taking account of eqn. (3.19). The minus sign indicates that R_Q acts downward in this instance.

Equations similar to (3.18) and (3.19) may be obtained for arches and rigidly jointed portal frameworks.

(*a*)

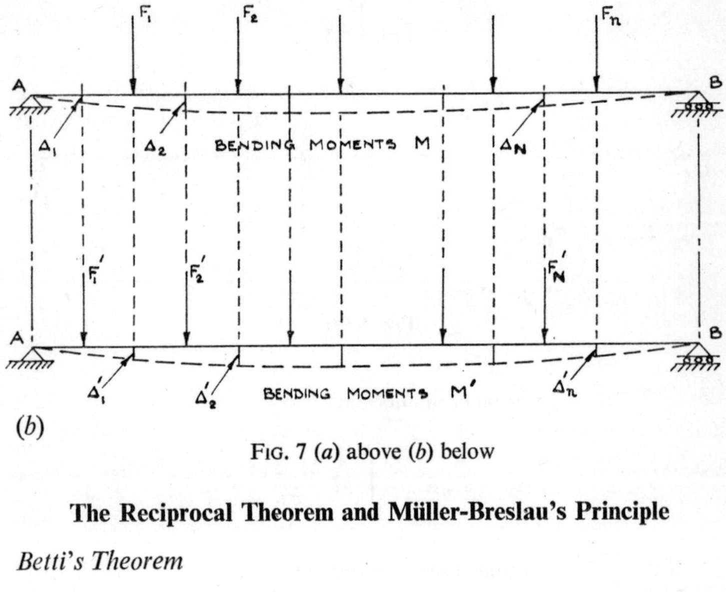

(*b*)

FIG. 7 (*a*) above (*b*) below

The Reciprocal Theorem and Müller-Breslau's Principle

Betti's Theorem

The principle of virtual work can be used to derive the form of the reciprocal theorem for linear systems, due to E. Betti, which is as follows:

$$F_1\Delta'_1 + F_2\Delta'_2 + \dots + F_n\Delta'_n = F'_1\Delta_1 + F'_2\Delta_2 + \dots + F'_n\Delta_N, \tag{3.21}$$

where $F_1, F_2, \dots, F_n$ and $F'_1, F'_2, \dots, F'_n$ are force systems applied separately to a linear elastic system and $\Delta_1, \Delta_2, \dots, \Delta_N$ are the deflections caused by the former system in the lines of action of the forces of the latter system, respectively; while $\Delta'_1, \Delta'_2, \dots, \Delta'_n$

are the deflections caused by the latter system in the lines of action of the forces of the former system, respectively. This is illustrated in Fig. 7 with reference to a beam as representing linear elastic systems generally.

Applying the principle of virtual work to the forces and bending moments, in equilibrium, shown in Fig. 7(*a*) using small virtual displacements equal to those caused by the force system of Fig. 7(*b*) gives:

$$\sum^{n} F_i \Delta_i' = \int_0^l M \frac{M' \, dx}{EI}. \tag{3.22}$$

But, application of the principle of virtual work to the forces and bending moments, in equilibrium, shown in Fig. 7(*b*) using small virtual displacements equal to those caused by the force system of Fig. 7(*a*) gives:

$$\sum^{N} F_j' \Delta_j = \int_0^l M' \frac{M \, dx}{EI}. \tag{3.23}$$

Therefore $$\sum^{n} F_i \Delta_i' = \sum^{N} F_j' \Delta_j \qquad \text{(eqn. 3.21)}$$

The reciprocal property of the mutual flexibility coefficients of linear elastic systems also follows from Betti's theorem. For example, if *P* and *Q* are two points on a beam, as shown in Fig. 8,

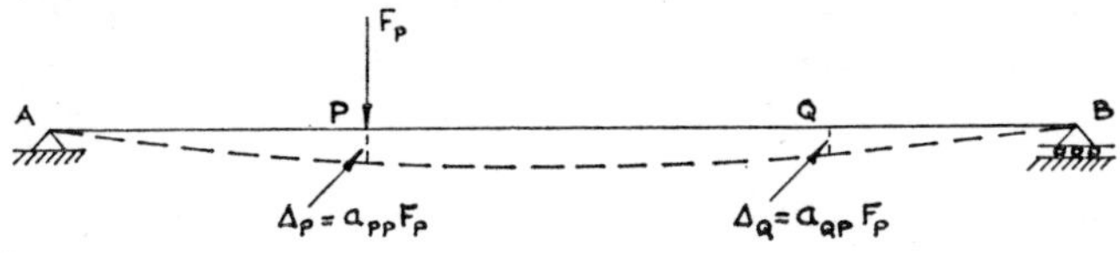

FIG. 8 (*a*)

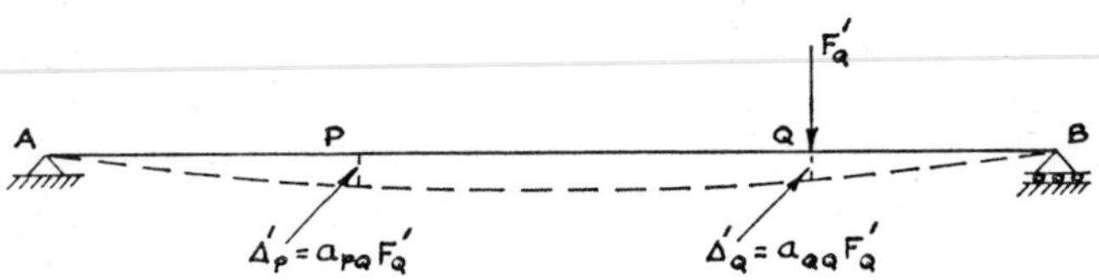

FIG. 8 (*b*)

then consideration of linear elastic behaviour for a load F_P at $P(F_Q = 0)$ yields:

$$\begin{aligned} \Delta_P &= a_{PP}F_P \\ \Delta_Q &= a_{QP}F_P, \end{aligned} \tag{3.24}$$

where a_{PP} and a_{PQ} are flexibility coefficients defined as the deflection of P due to unit load at P and deflection of Q due to unit load at P, respectively. Thus, the first letter of the subscripts denotes the point at which the deflection is being considered while the second letter denotes the point of application of the load which causes the deflection. Again, for a load F'_Q at $Q(F'_P = 0)$:

$$\begin{aligned} \Delta'_P &= a_{PQ}F'_Q \\ \Delta'_Q &= a_{QQ}F'_Q. \end{aligned} \tag{3.25}$$

Application of Betti's theorem (eqn. 3.21) to the two systems of forces and deflections, now, gives:

$$F_P\Delta'_P + O\Delta'_Q = O\Delta_P + F'_Q\Delta_Q, \tag{3.26}$$

or

$$F_PF'_Qa_{PQ} = F'_QF_Pa_{QP}, \tag{3.27}$$

that is:

$$a_{PQ} = a_{QP}, \tag{3.28}$$

so verifying the reciprocal property of mutual flexibility coefficients. The "self-coefficients" such as a_{PP} and a_{QQ} are, of course, independent.

Müller–Breslau's Principle by the Concept of Flexibility Coefficients

The concept of flexibility coefficients can be used directly to justify Müller-Breslau's principle for linear statically indeterminate systems having regard to the reciprocal properties of these coefficients. With reference to the continuous beam (Fig. 6), for example, in the absence of a support at Q, the deflection there due to load F at X would be $a_{QX}F$ upward, if a_{QX} is defined as the upward deflection of Q due to unit load at X. Thus, when Q is supported, the support must provide at any time just sufficient force to nullify the deflection which any loading would cause at Q in the absence of the support. That is, if R_Q is the support reaction upward due to load F at X:

$$\Delta_Q = a_{QQ}R_Q + a_{QX}F = 0, \tag{3.29}$$

$$R_Q = -\frac{a_{QX}}{a_{QQ}}F, \tag{3.30}$$

that is, R_Q acts downward.

Now, the introduction of a displacement Δ'_Q at Q by an arbitrary force R'_Q causes a vertical deflection Δ'_X (positive upward) at any point X such that:

$$\Delta'_X = -R'_Q a_{XQ}, \tag{3.31}$$

but

$$\Delta'_Q = a_{QQ}R'_Q, \tag{3.32}$$

therefore:

$$\Delta'_X = -\frac{a_{XQ}}{a_{QQ}}\Delta'_Q, \tag{3.33}$$

or

$$\frac{a_{XQ}}{a_{QQ}} = \frac{a_{QX}}{a_{QQ}} = -\frac{\Delta'_X}{\Delta'_Q}, \tag{3.34}$$

since $a_{XQ} = a_{QX}$ by the reciprocal theorem. Substitution in eqn. (3.30) then gives:

$$R_Q = -\frac{\Delta'_X}{\Delta'_Q}F, \tag{3.35}$$

for a downward load F at any point X. In other words, the displaced or deflected form of the beam due to the introduction of a small displacement Δ'_Q in the line of action of R_Q gives the influence line for R_Q.

Müller–Breslau's Principle by Direct Application of Betti's Theorem

The linearly elastic arch shown in Fig. 9 is appropriate for illustrating direct application of Betti's theorem to verify Müller–Breslau's principle. It is encastré in rigid abutments at A and B.

The arch is statically indeterminate due to three redundants, for example, the three components of reaction provided by the abutment at A. By Müller–Breslau's principle the influence line for the horizontal component R_1 of the reaction at abutment A is represented by the displaced form of the arch which is compatible with the introduction of a small displacement Δ_1', at A, as shown in Fig. 9(b). The components of the abutment reactions due to a

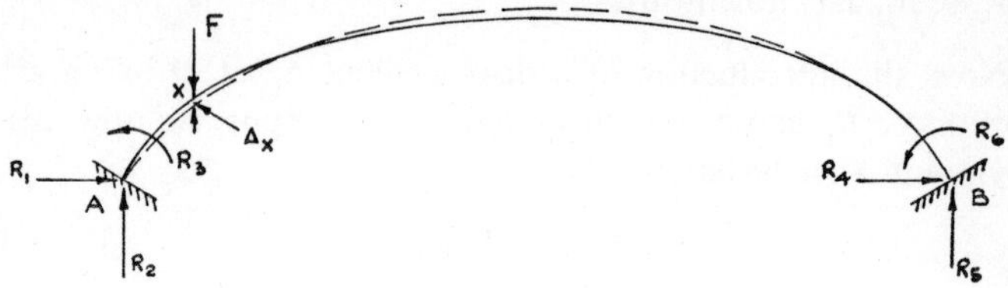

FIG. 9 (a)

FIG. 9 (b)

concentrated load F at X are shown in Fig. 9(a) with the deflected form of the arch caused by this load. The forces R_1' and R_2' and the couple R_3' which are necessary to produce a small displacement Δ_1', at A, of the arch supported rigidly at B are shown in Fig. 9(b). Application of Betti's theorem to the systems of forces and displacements shown in Figs. 9(a) and (b) gives:

$$R_1\Delta_1' + R_2 O + R_3 O - F\Delta_X' = R_1' O + R_2' O + R_3' O, \quad (3.36)$$

so that,

$$R_1 = \frac{\Delta_X'}{\Delta_1'} F, \quad (3.37)$$

which is in accordance with Müller–Breslau's principle. A similar procedure may be used to verify that the influence lines for R_2, the vertical component of the reaction at abutment A and R_3 the reactive couple at A are represented by the displaced forms of the arch which are compatible with small displacements Δ_2' vertically and Δ_3' rotationally, respectively, of A.

Equation (3.37) may be derived by direct application of the principle of virtual work as for the derivation of eqn. (3.20) above, for a continuous beam, if deformation of the arch due to bending only is considered. Application of Betti's theorem to the continuous beam problem leads, of course, to eqn. (3.20).

Influence Lines for Deflection

The influence line for the deflection of any point of a linear elastic system in any direction is directly proportional to the deflected form of the system due to a concentrated load applied at the point and acting in the specified direction. This may be verified by considering a simply-supported beam whose elasticity is linear, as shown in Fig. 8.

Since for any two points on the beam $a_{PQ} = a_{QP}$ by the reciprocal theorem, a load F_Q' applied at Q will cause a deflection $\Delta_P' = a_{PQ}F_Q'$ at P so that $a_{PQ} = \Delta_P'/F_Q' = a_{QP}$, the deflection at Q due to unit load at P. Hence, the deflected form shown in Fig. 8(*b*) due to a vertical load F_Q' at Q represents the influence line for the vertical deflection of Q. In fact, the influence line for the deflection of Q is precisely the deformed shape of the beam due to unit load at Q. Further generalization is unnecessary, a simple beam with linear elasticity is representative of any structure, however complex, which behaves linearly. Also, this theorem, which has been attributed to O. Mohr, is the basis of the moment deformeter whereby the influence line for the curvature at part of a member of a framework may be obtained.

The principle of the moment deformeter may be appreciated by considering the beam system with linear elasticity shown in Fig. 10(*a*). A load F_Q is applied at any point Q, causes a deflection

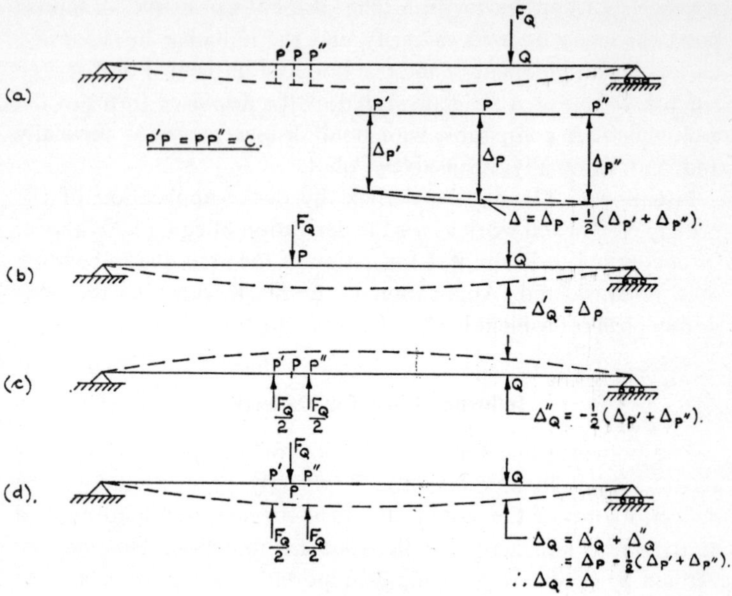

FIG. 10

Δ_P at any other point P and deflections $\Delta_{P'}$ and $\Delta_{P''}$ at points P' and P'' equidistant from P by a small amount "C", as shown in Fig. 10(*a*). Assuming that the deformed length $P'PP''$ of the beam is circular of radius R_P, then by the property of a circular arc:

$$(2R_P - \Delta)\Delta = C^2, \tag{3.38}$$

where Δ is defined in Fig. 10(*a*). Treating Δ^2 as being negligible, therefore, gives:

$$\frac{1}{R_P} = \frac{2\Delta}{C^2} \tag{3.39}$$

That is, the curvature is directly proportional to Δ. Now, superposition of the systems of loading shown in Fig. 10(*b*) and (*c*) as shown in Fig. 10(*d*) causes a deflection of any point Q outside the region $P'P''$:

$$\Delta_Q = \Delta_P - \tfrac{1}{2}(\Delta_{P'} + \Delta_{P''}) = \Delta, \tag{3.40}$$

having regard to the reciprocal theorem. That is, the system of loads shown in Fig. 10(*d*) causes the beam system to deform in such a way that the deformed shape is directly proportional to the influence line for curvature at *P*. The moment deformeter is merely a device for introducing the system of loads shown in Fig. 10(*d*) to a scale model.

Members in Bending and Treatment of Loading in Relation to Model Analysis

Slope–Deflection Relationships

When an initially straight uniform beam *AB* with linear elasticity is simply supported and subjected to a terminal couple M_A,

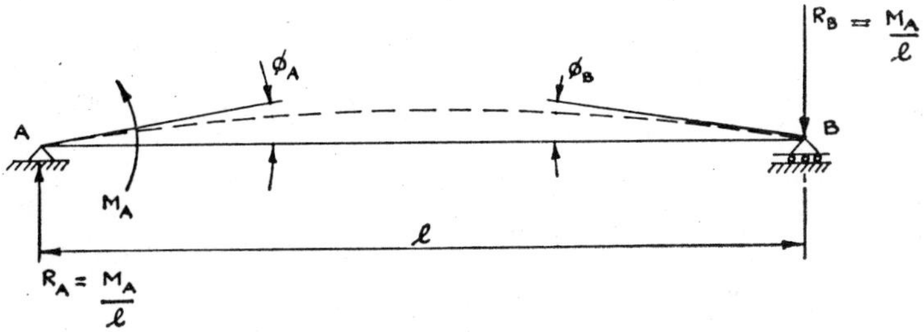

FIG. 11

as shown in Fig. 11, reactions at the supports of M_A/l are necessary for equilibrium, as shown, and the slopes at the ends of the beam due to M_A are:

$$\phi_A = \frac{l}{3EI} M_A : \phi_B = -\frac{l}{6EI} M_A. \tag{3.41}$$

The sign convention here is in accordance with the usual convention for Cartesian co-ordinates, where couples and rotations are positive when they are in the anti-clockwise sense. Similarly, for a positive terminal couple M_B:

$$\phi_A = -\frac{l}{6EI} M_B : \phi_B = \frac{l}{3EI} M_B, \tag{3.42}$$

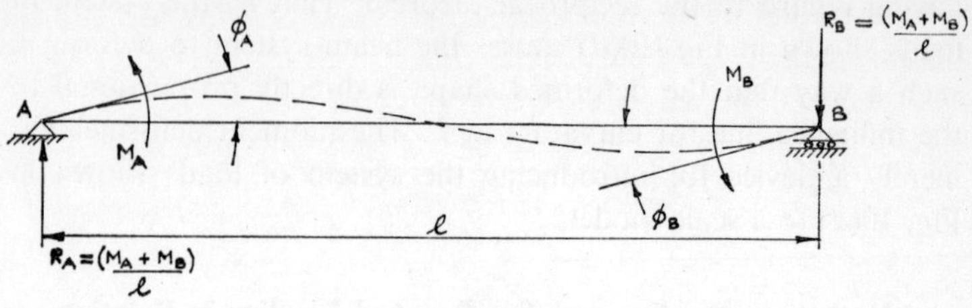

FIG. 12

while for positive terminal couples M_A and M_B applied simultaneously, necessitating support reactions $(M_A + M_B)/l$ as shown in Fig. 12:

$$\phi_A = \frac{l}{3EI} M_A - \frac{l}{6EI} M_B = \frac{l}{6EI}(2M_A - M_B)$$

$$\phi_B = -\frac{l}{6EI} M_A + \frac{l}{3EI} M_B = \frac{l}{6EI}(-M_A + 2M_B). \quad (3.43)$$

Alternatively, these equations may be inverted to give:

$$M_A = \frac{2EI}{l}(2\phi_A + \phi_B)$$

and (3.44)

$$M_B = \frac{2EI}{l}(\phi_A + 2\phi_B).$$

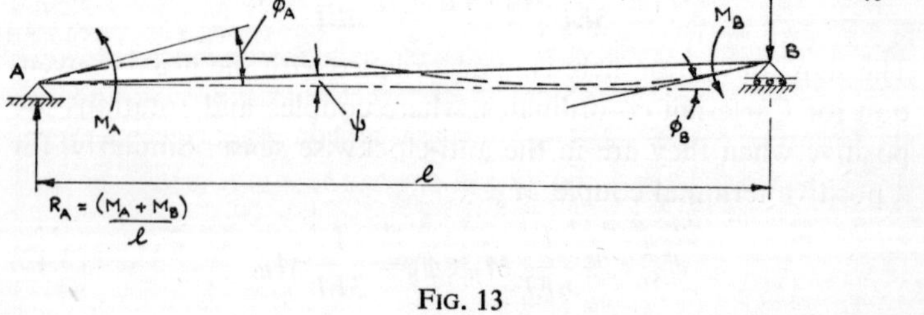

FIG. 13

In the event of the line joining the supports being inclined to the x-axis by a small positive angle ψ, as shown in Fig. 13 eqns. (3.43) and (3.44) become, respectively:

$$\begin{aligned} \phi_A - \psi &= \frac{l}{6EI}(2M_A - M_B) \\ \phi_B - \psi &= \frac{l}{6EI}(-M_A + 2M_B) \end{aligned} \tag{3.45}$$

and

$$\begin{aligned} M_A &= \frac{2EI}{l}[2(\phi_A - \psi) + (\phi_B - \psi)] \\ M_B &= \frac{2EI}{l}[(\phi_A - \psi) + 2(\phi_B - \psi)], \end{aligned} \tag{3.46}$$

if ϕ_A and ϕ_B are measured from the x-axis.

Equations (3.44) are the basis of a simple but powerful instrument for the model analysis of structures, called the "moment indicator" which is described in Chapter 6.

Treatment of Loads on Beams in Structures

For the purpose of structural analysis it is sometimes desirable to consider all loads as being applied at joints or intersections of members. Having analysed a structure on this basis and so determined the forces applied to the ends of members, it is then necessary to consider the individual members under the influence of their actual load distribution in equilibrium with the terminal forces.

This approach to loading could be useful when considering the loading of models by compensating balance, for direct tests. There is, however, the difficulty of applying the necessary equivalent load couples at the joints of models of rigidly jointed frameworks which would usually be involved, though in some instances they would be negligible due to equal and opposite effects from adjacent members (for example, in rigidly jointed building frameworks subjected to wind loading).

If a beam *AB* as part of a structure is loaded as shown in Fig. 14(*a*) then it will have no effect on the structure as a whole if external forces and couples of appropriate magnitude are applied at the joints *A* and *B* as shown in Fig 14(*b*). It follows, then, that as far as the structure as a whole is concerned, the effect of the loading within the beam *AB* is the same as if the system of "restraining" forces and couples shown in Fig. 14(*b*) is reversed

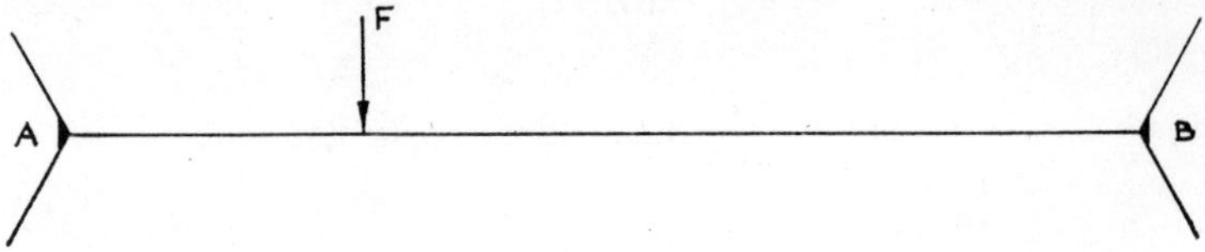

FIG. 14 (*a*)

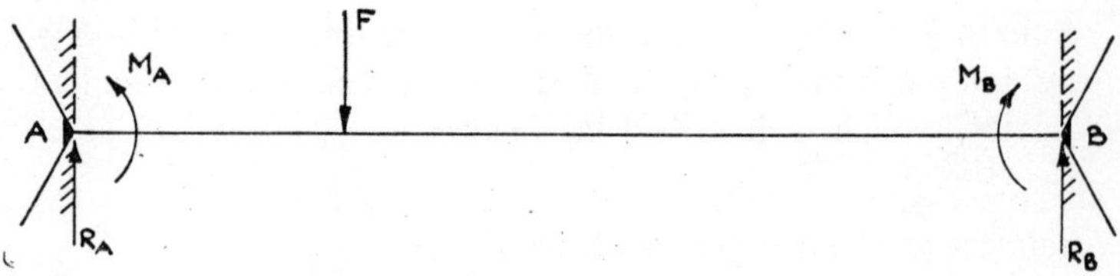

FIG. 14 (*b*)

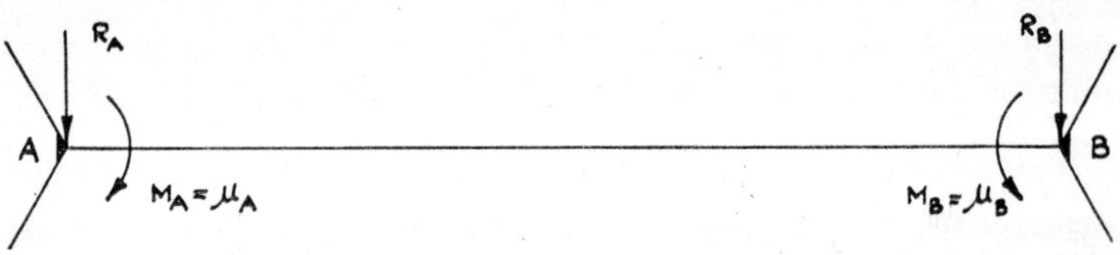

FIG. 14 (*c*)

and applied at *A* and *B* as shown in Fig. 14(*c*). In other words, superposition of the external force systems of Figs. 14(*b*) and (*c*) gives the external force system of Fig. 14(*a*).

The magnitude of the restraining forces and couples is clearly identical to the supporting forces and couples which would be necessary to make the loaded beam encastré. The couples are, therefore, the "fixed-end" moments for the beam, while the restraining forces are such that together they are equal in magnitude but opposite in sense to the beam loading, their individual values being determined by the requirement of zero resultant couple on the beam as a whole. Values of fixed-end moments and reactions for three components of load on a uniform beam with linear elasticity are shown in Fig. 15(*a*), (*b*) and (*c*).

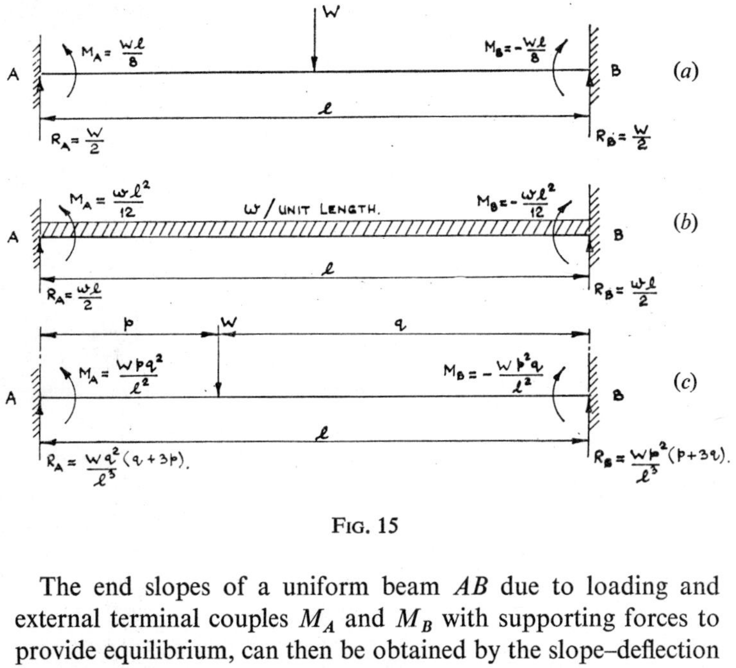

FIG. 15

The end slopes of a uniform beam *AB* due to loading and external terminal couples M_A and M_B with supporting forces to provide equilibrium, can then be obtained by the slope–deflection equations, eqn. (3.43). However, M_A and M_B must be replaced by $(M_A + \mu_A)$ and $(M_B + \mu_B)$, respectively, where μ_A and μ_B denote the reversed fixed-end moments to take account of the loading within the span of the beam.

CHAPTER 4

MODELS OF PLANE FRAMEWORKS

Nature of Models

Models of frameworks are scaled-down versions of their respective prototypes, but, as described in Chapter 2, the conditions for flexural similarity can often be fulfilled for plane structures, without reproducing the cross-sectional detail of the prototype members on the model. Models of portal frames can be cut from sheets of celluloid or perspex. At first sight they might appear to be exact geometrical replicas of their prototypes, but closer examination reveals that all of the members have rectangular cross-sections, the breadth of which is fixed by the thickness of the sheet material. Joints within members for enabling internal displacements to be applied to obtain influence lines by Müller–Breslau's principle, also make a model different in appearance from its prototype (see Fig. 16).

Models of trusses can be made of metal rod or strip with flexible elements within members to provide the requisite axial flexibility as shown in Fig. 17.

Mounting of models on a vertical baseboard overcomes the problem of contact friction between model and baseboard to a large extent. Provision may be made in the mounting arrangements to enable displacements to be introduced at supports for the purpose of obtaining influence lines for support reactions.

General Design of Models

Apart from the requirements of flexural similarity to the prototype which are considered in Chapter 2, a model must be designed

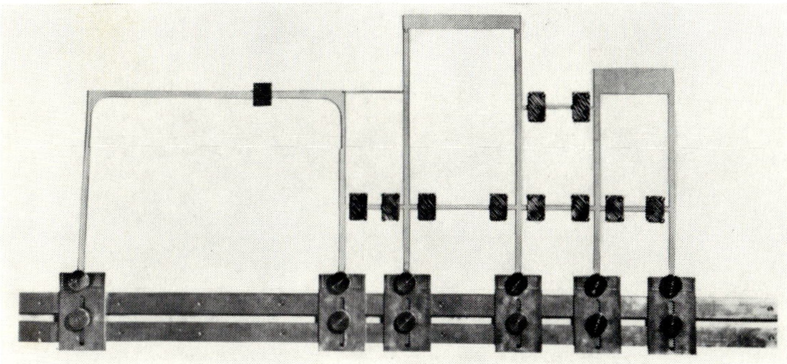

FIG. 16

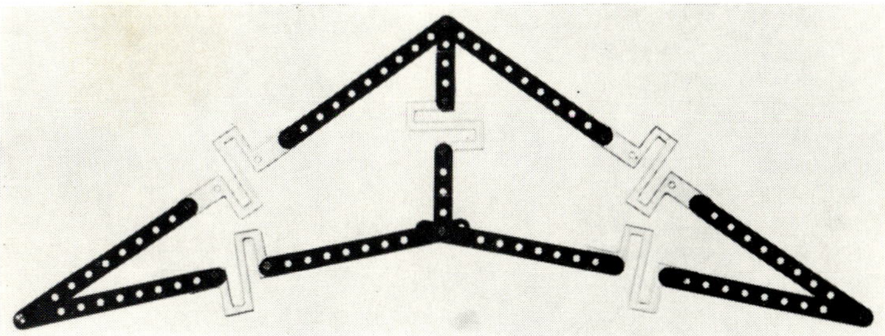

FIG. 17

(*Facing p.* 40)

so that it can be used readily to give accurate results. The choice of scale factors is fundamental in this respect, because the model should be sufficiently flexible for the requisite displacements or deformations to be introduced without exceeding the elastic limit at any point, or causing buckling or gross distortion. The magnitude of the displacements must be such that the ordinates of the resulting deflected forms (which represent influence lines) or moment indicator movements can be measured accurately with the naked eye.

Design of Models of Portal Frames

For most of the problems which are conveniently analysed by model test, the deflections due to bending of members in the plane of the framework only are important and suitable models can be cut from sheet material of uniform thickness. (This does not mean that axial and shear forces cannot be found by using the model, as is explained later.) The selection of the thickness of material, having regard to the scale factor which has been chosen for the layout dimensions of the model, requires careful consideration. If the material is too thin in relation to the size of the model, it will rest against the baseboard as well as tend to buckle when displacements are introduced. For most applications $\frac{3}{16}$ in. thick material is satisfactory, though for models of simple frameworks $\frac{1}{8}$ in. thick material will probably suffice.

For reasons of model manufacture it is desirable that the depth of the most slender member is not less than approximately $\frac{1}{8}$ in., otherwise the scale factor for the cross-sections of the members can be chosen freely without reference to the layout scale factor. In general, the layout scale should be as large as practicable and the depth of the most slender member as small as practicable.

For a given thickness of material it is a simple matter to determine the depths of the members for a chosen second moment of area scale factor. If the relevant second moment of area of a member of the prototype structure is I and a scale factor r (model dimensions $1/r$ times those of the prototype) has been selected for cross-section dimensions, then for the corresponding model member:

$$I' = \frac{I}{r^4}. \tag{4.1}$$

Thus, the ratios of the second moments of area of the members of the model will be the same as those of members of the prototype, respectively, being the requirement for flexural similarity when axial and shear deformations are negligible. Since the model member cut from sheet will have a rectangular cross-section of thickness, say t, the depth d will be given by:

$$d = \sqrt[3]{\frac{12I'}{t}} = \sqrt[3]{\frac{12I}{r^4 t}} \tag{4.2}$$

because $I' = td^3/12$. Small errors in the depth of members (d) may not therefore be negligible. The design of joints within members to enable internal displacements to be applied is considered below.

At supports a generous allowance of spare material should be made on the model for fixing to the baseboard and its associated apparatus. This provision is not, however, usually necessary for models of bridge frames or girders, as is shown in Fig. 26.

Design of Models of Trusses

The design of models of trusses, which are such that it is justifiable to assume that members are subject to axial forces only, is almost entirely a question of designing suitable flexible elements for inclusion within the members. The flexible elements are necessary to overcome the inherent axial rigidity of members made of metal or celluloid strip. It is usually correct to assume that members with flexible elements derive the whole of their axial flexibility therefrom. The design of the elements is considered below and is based upon the requirement that the ratios of the axial flexibilities of the model members must be the same as the corresponding ratios for the prototype. The model members as constructed with flexible elements must, of course, possess sufficient lateral rigidity to avoid buckling when displacements or loads are introduced into the model in accordance with the technique of testing. Means of introducing internal displacements can be arranged conveniently at the flexible elements.

A model of a truss made of Meccano and celluloid flexible elements is shown in Fig. 17.

Model Materials

Model materials must possess the same elastic properties as those of the respective prototype structures. Thus for linear structures, the model material must be homogeneous and obey Hooke's law. Sometime it is necessary, in addition, for the Poisson's ratio of the model material to be the same as that of the prototype, but this factor can usually be disregarded for plane structures.

The second important factor in the choice of model materials is workability. Thus it is desirable that models can be made by means of simple hand-tools suitable for use in the design-office. A few useful and common materials are considered below, and there are, no doubt, a number of alternatives which are equally satisfactory.

Xylonite-celluloid, Bakelite and Perspex

Xylonite-celluloid is possibly the most generally suitable material for models of portal frames. It is tough and easy to cut and has a linear stress–strain characteristic with a low modulus of elasticity. Celluloid is rather sensitive to atmospheric humidity and unless seasoned sheets are used a certain amount of warping might occur but would not usually be sufficient to be troublesome in model testing. The creep characteristics of celluloid are such that compensation for creep may be achieved easily as indicated below and in Chapter 6.

Possible alternatives to celluloid are photoelastic bakelite and perspex. The former has properties similar to those of celluloid but perspex is more brittle and has a higher modulus of elasticity which renders it generally unsuitable for all except the simpler models when large displacements are to be introduced. Moreover, it is not strictly linearly elastic. On the other hand its creep rate is sufficiently low to enable it to be used for direct testing of models of complex structures by the application of dead loads.

These materials are manufactured commercially in sheets of

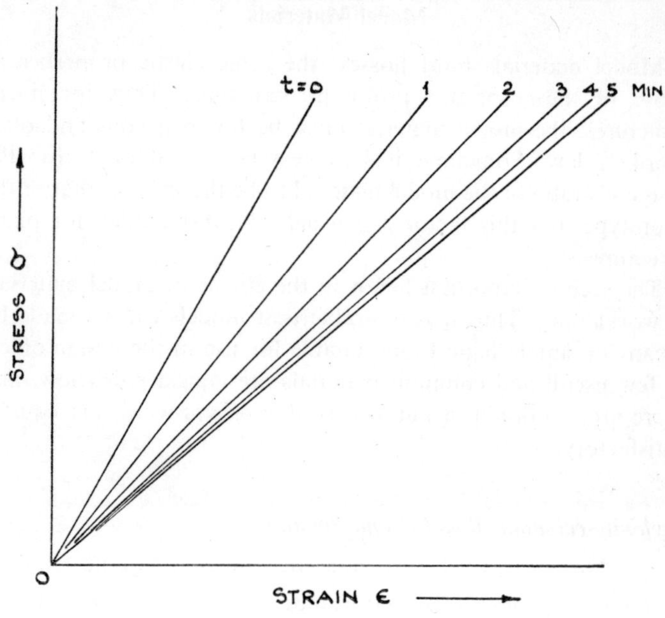

FIG. 18

various thicknesses, and $\frac{3}{16}$ in. thick material is satisfactory for most models. It is important to check the thickness of the sheets at a number of points, using a micrometer gauge, before making a model. The variation in thickness should not be greater than a few thousandths of an inch unless suitable corrections are made to the cross-sectional dimensions of the model members.

Celluloid and bakelite creep when they are loaded such that

their Young's moduli appear to change with time. Thus, the straight lines in Fig. 18 represent the loci of the strains produced by constant stresses in a specimen of one of the materials as time elapses. For a stress defined by a horizontal line drawn through the end points of the lines and applied at time $t = 0$ the instantaneous strain is represented by point O, while points 1, 2, 3, etc., represent the strains at $t = 1, 2, 3$ min respectively, after application of the

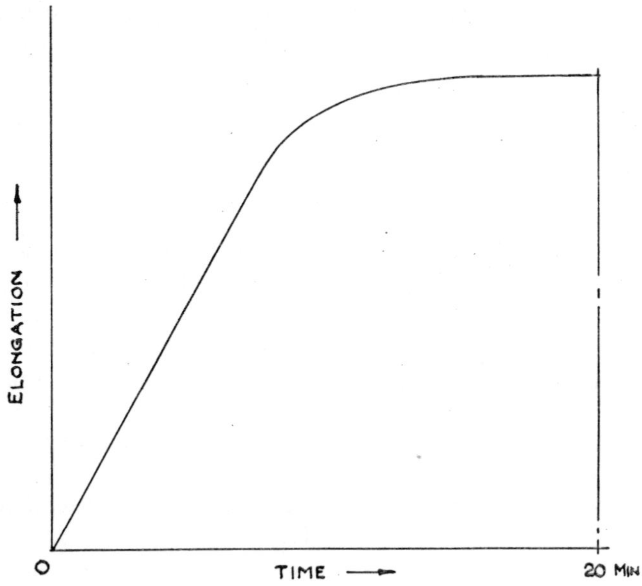

FIG. 19

stress. The instantaneous strain due to a smaller stress applied at $t = 0$ would lie on the line O, and at $t = 1, 2, 3$ min, the strains due to this stress would be determined by the appropriate points on lines 1, 2, 3, etc. If unloading takes place instantaneously the strain decreases initially along a line parallel to O, and the residual strain disappears gradually over a period of time. Figure 19 illustrates the nature of the strain due to creep with respect to time.

By virtue of this creep characteristic, a fixed deflection introduced in a celluloid model (as opposed to a constant force) causes a deformed shape which does not vary due to creep, though the force necessary to maintain the fixed deflection does vary.

Celluloid, bakelite and perspex models may be repaired by means of the appropriate adhesive cements. (Celluloid is soluble in acetone while perspex is soluble in chloroform.) Cemented joints are not, however, recommended for the manufacture of models. Wherever possible, cemented joints should be such that the adhesive has to withstand shearing action only. Araldite cement is, incidentally, a strong adhesive which is suitable for a wide range of materials.

Metal

Models of linear structures of the portal type can be made from brass or steel strip and suitable standard clamping and connecting pieces in the manner advocated by Gottschalk and Rieckhof. It is, however, almost impossible to have available a sufficiently wide range of strips of various thicknesses to enable a model of any given prototype to be made so that it is truly flexurally similar. The alternative of cutting models from sheet metal is unnecessary having regard to the availability of easily worked sheet plastic materials.

Steel or brass rods or strips are suitable for models of truss structures since the requisite relative elastic properties of the members can be obtained by making them in two parts connected by suitable flexible elements. Meccano strip should always be considered as being a readily available material. It is important to note, however, that when the members of a model of a truss have celluloid flexible elements the above remarks on the liability of models to creep under constant load are applicable.

Cardboard

By virtue of the uncertain elastic properties of cardboard, it is generally suitable only for rapid preliminary model investigations for relatively simple structures, e.g. single-bay portal frames with non-uniform members. It is important that good-quality cardboard is used in order to lessen the chances of poor results due to variable elastic properties.

Manufacture of Models

Models of portal frames may be cut from uniformly thick sheets of celluloid or perspex by means of a fret-saw. A model is carefully marked out on the sheet of material with a sharp scriber, and a small allowance made during cutting out so that a finished width of members accurate to within a few thousandths of an inch can be obtained by filing. Provision must be made for mounting the models, for example, substantial pieces of material or lugs must be allowed at the feet of columns of portal frameworks which are to be encastré. In order to prevent stress concentrations and possible breakage of a model when it is deformed, sharp corners at junctions of members must be avoided during manufacture. Standard metal pins and accessories which are sometimes necessary for fixing the model to a mounting or for the joints within members which are described below, can be purchased (e.g. hard steel pins of the kind used for hanging pictures) or made once for all.

Metal (or a combination of metal and celluloid as shown in Fig. 17) is probably the most suitable material for models of trusses whose deflection is due primarily to axial deformation of members.

Mounting of Models

Models may be mounted for testing by means of sturdy metal pins or otherwise on a vertical baseboard (Fig. 16). They should not actually touch the baseboard, which is merely for the purpose of enabling deflections or deformed shapes of models to be recorded on paper otherwise contact friction will be a source of error. While it is possible that celluloid models, mounted vertically, will creep due to their own weight, the effect is usually negligible for practical purposes.

A horizontal baseboard may be used, if desired, but then models must be supported on ball-bearings at a number of points to eliminate contact friction.

Introduction of Displacements at Supports

If a model is fixed to the baseboard by metal pins the introduction of displacements at supports is easy to achieve but this method of mounting is not generally suitable for models of complicated portal frameworks. Accordingly more elaborate mounting arrangements may be considered as described below.

By one method of mounting models of portal frameworks, the feet or supports of the models are each drilled and clamped by screws to a rigid metal plate which is attached to a slider in a slot along the lower edge of a baseboard. Each plate has a vertical slot at the screw attachment to its respective slider to enable vertical displacements to be introduced. Horizontal displacements are introduced merely by unclamping and moving the sliders. The same screw can be used for clamping both the plate to the slider and the slider in the horizontal slot. These details are shown diagrammatically in Fig. 20. Rotational displacements may be introduced at the clamping screws of the feet of the models.

While this method of supporting and introducing displacements to models is effective and adaptable, a more refined design has a "fixed" plate attached to each slider, to which the plate fixed to the model foot is connected by fitted pins in bushed holes. By having different sets of holes in the "fixed" plates into which the connecting pins can be inserted, either horizontal, vertical or rotational displacements can be introduced at the model feet. This method of introducing the displacements is illustrated in Fig. 21.

The simple method of mounting by metal pins driven into the baseboard is usually satisfactory for models of simply supported or continuous trusses and bridge girders. It is then important to ensure that longitudinal freedom is permitted by providing slots

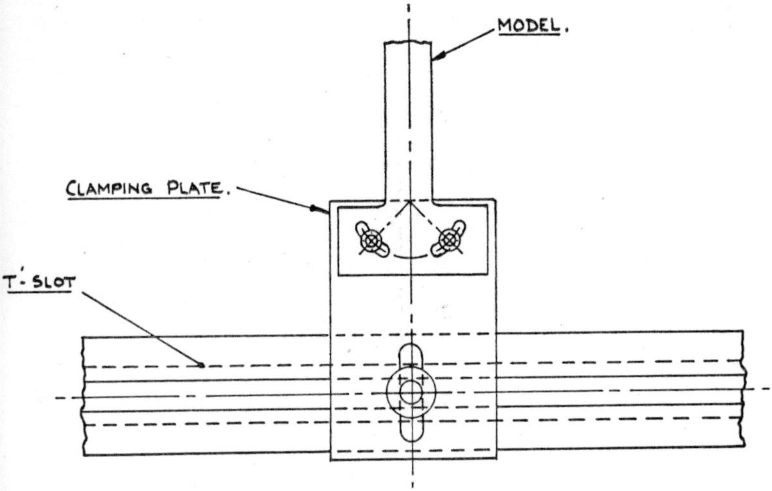

FIG. 20

in the model for mounting pins which represent supports for providing vertical restraint only.

Introduction of Internal Displacements

The method of introducing internal displacements is by means of internal joints. While it is necessary to provide for axial displacements only in the members of models of trusses, provision must usually be made for axial, shear and rotational (bending) displacements within the members of models of portal frameworks (an exception is those of the single-storey type).

The connections for the flexible elements which are incorporated in members of models of redundant trusses can be used to serve the purpose of internal joints. Thus, a displacement within a member can be introduced by disconnecting the flexible element and reconnecting in such a way that the length of the member is altered. It is merely necessary to provide alternative holes for bolting the flexible elements within members.

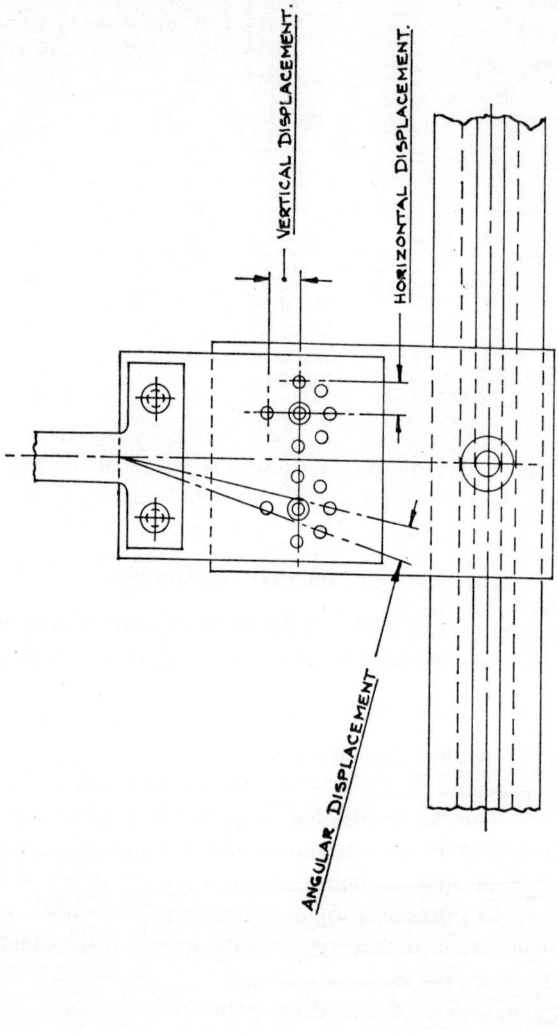
VERTICAL DISPLACEMENT.
HORIZONTAL DISPLACEMENT.
ANGULAR DISPLACEMENT

FIG. 21

As the deformation of portal frames is due primarily to the bending of members the joints for allowing displacements to be introduced must be such that the relevant flexural properties of the members are retained. The type of joint shown in Fig. 22(*a*) may be used. Adjacent ends of a "broken" member are provided with transverse bars, the connection being made by metal plates

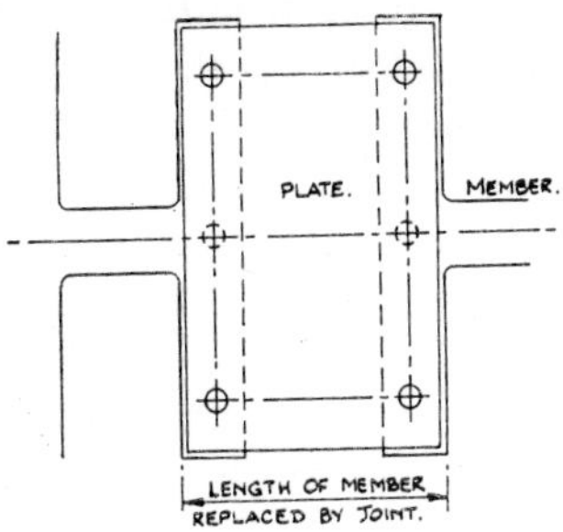

FIG. 22 (*a*)

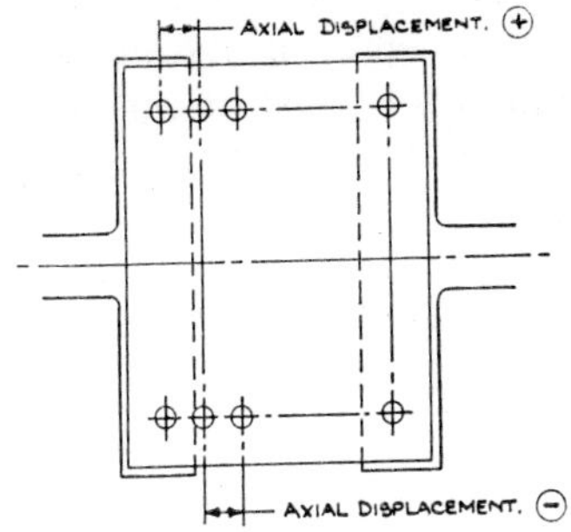

FIG. 22 (*b*)

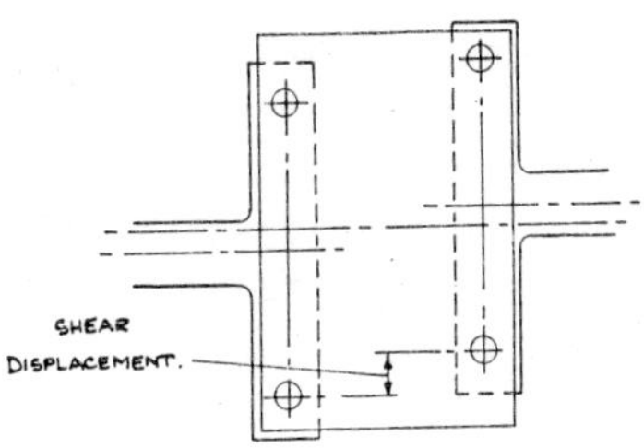

FIG. 22 (*c*)

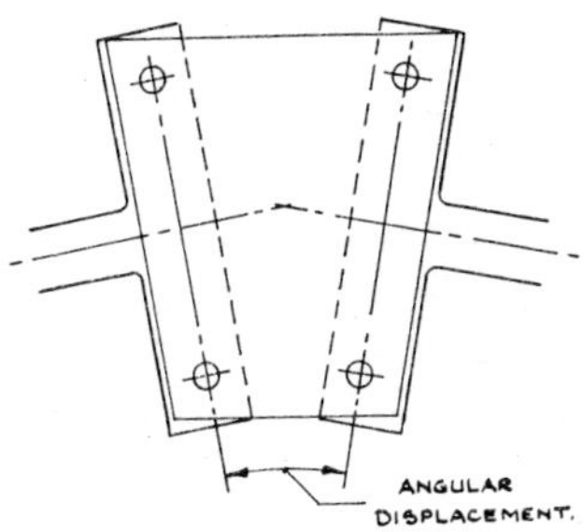

FIG. 22 (*d*)

fixed to the bars by fitted pins. The proportions of the bars are such that the flexibility of a joint in bending is the same as the length of member which it replaces.

Axial, shear and rotational displacements may be introduced by replacing the normal connecting plates by similar plates with the holes drilled to suit the magnitude and sense of the respective displacements as shown in Fig. 22(*b*), (*c*) and (*d*). Thus, for an axial displacement the distance between the vertical centre lines of the pairs of holes in a plate must be different from the normal by the amount of the displacement. Again, for a shear displacement, the pairs of holes must be offset by the amount of the displacement, while for a rotational displacement the pairs of holes must be drilled on converging centre lines, the angle between which represents the magnitude of the displacement. It is essential that the normal spacing of the joint on the axis of the member is retained when a rotational displacement is applied.

It should be noted that the joints will generally modify the axial and shear flexibilities of the members, and it is theoretically possible for errors to accrue from this source. In order to ensure that the axial flexibility of a member is not increased significantly by a joint, however, additional pins can be used at the centre line of the member, for securing the joint plates, as shown in Fig. 22(*a*).

Design of Internal Joints and Flexible Elements

The design of the joints within members of models of portal frameworks for enabling internal displacements to be introduced and the design of flexible elements for members of statically indeterminate trusses is based upon the achievement of the requisite flexibility. Care must, however, be taken to ensure that the proportions of these items are not unduly large in relation to the layout dimensions of models to the extent that interference is caused when loading or displacements are introduced during testing of the models.

A flexible element in the form of a celluloid ring or "*S*" piece, provides the connection between the two parts of a truss member. It may be bolted to the adjacent ends of the two parts of a member as shown in Fig. 17, extra bolt holes being provided at appropriate spacing to enable an axial displacement to be introduced into the member. If the overall proportions of all flexible elements of a model are identical, the required ratios of axial flexibilities of members may be obtained simply by differences in depths of the parts of the elements. Thus, the flexibilities of the joints must be in the same ratio, one to another, as the ratios of the axial flexibilities of the corresponding members of the prototype. If, then, the uniform depth of the flexible element of one member is d_1, while that of another member is d_2, the ratio of the flexibilities of the members $a_1 : a_2$ is given by the ratio $d_1^3 : d_2^3$ provided that the thickness of material of both is the same. This is because the flexibility of the elements depends upon the flexibility in bending of their parts which is inversely proportional to the second moment of area of those parts.

For portal frameworks, the transverse bars which are formed at the adjacent ends of the members for the purpose of joints are proportioned so that the bending flexibility of the members is unchanged by the joints. Referring to Fig. 23, if the length of member of flexural rigidity $E'I'$ which is replaced by a joint is λ, and a bending moment M' is applied to this length, the resulting change of slope ϕ between its ends is given by:

$$\phi = \frac{M'\lambda}{E'I'}, \tag{4.3}$$

by the Bernoulli–Euler theory, which incidentally, is only approximate for short, deep beams of which this is an example. The proportions of the transverse bars must, therefore, be such that when the moment M' is applied to the joint, the resulting change of slope is again ϕ. If each bar is considered to consist of two cantilevers of length $(l_c - d)/2$ where l_c is the distance between the holes in the bars for receiving the connecting pins, the change of slope over the joint is:

$$\phi = \frac{4\Delta_c}{l_c}, \tag{4.4}$$

where Δ_c is the deflection of each cantilever. For equivalence in bending, therefore, eqns. (4.3) and (4.4) must be identical, i.e.

$$\frac{4\Delta_c}{l_c} = \frac{M'\lambda}{E'I'}. \tag{4.5}$$

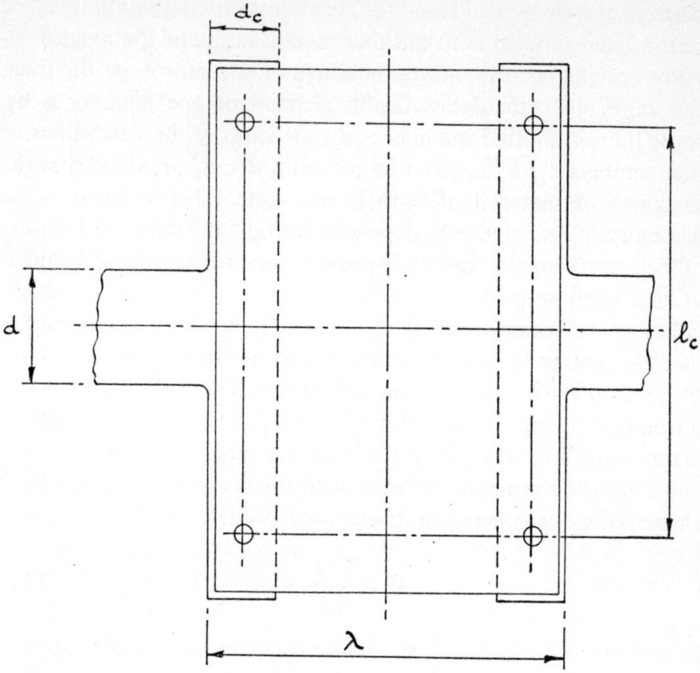

FIG. 23

Now $\Delta_c = F_c\,[(l_c - d)/2]\,^3/3E'I_c$ by the well-known expression for the deflection of a cantilever loaded at its end point, while $M' = F_c l_c$, $I' = td^3/12$ and $I_c = td_c^3/12$ where the subscript c denotes bar quantities, so that by eqn. (4.5):

$$\frac{1}{l_c^2}\left(\frac{l_c - d}{d_c}\right)^3 = \frac{6\lambda}{d^3}. \tag{4.6}$$

Of the quantities in eqn. (4.6) only d is fixed; it is therefore necessary to select values for any two of λ, l_c and d_c and determine the third by means of the equation. The dimensions λ should be as small as possible (consistent with the function of the joint) in the interests of true flexural similarity between the model and prototype.

As stated above, this method of designing internal joints is only approximate, and the underlying assumptions represent only one set of several alternatives. It may be improved by taking the deflection of the transverse bars due to their shear into account. It is, incidentally, often desirable that the joints are designed with a view to using standard connecting plates. Thus λ and l_c are predetermined and d_c must be calculated.

The connecting plates and pins, for the internal joints in celluloid or perspex models should be as light as possible to avoid noticeable deflection due to their weight when the models are mounted vertically. Connecting plates made of $\frac{1}{16}$ in. thick aluminium or brass sheet and $\frac{1}{16}$ in. diameter bright steel pins with knurled brass heads are generally suitable.

It is worth recalling here that Beggs practised the cutting of members, as required, for the introduction of displacements. Metal transverse bars were clamped to the member, one at each side of the cut, to which the equipment for introducing displacements was attached. The models were generally mounted horizontally, and no attempt was made to compensate for the change of flexural properties due to the joint, but the distance between the transverse bars was small because only small displacements were introduced.

It should be noted finally that if a model of a portal-type framework is to be tested using only the moment indicator or similar device, the provision of internal joints is unnecessary. Moreover, even for the purpose of obtaining influence lines for redundants, internal joints are essential only for portals having more than one storey or "closed" bays of the kind found in Vierendeel girders.

Typical Models of Rigidly Jointed Frameworks

Power Station Building Frame

A dimensioned line diagram of a transverse frame of a power station structure is shown in Fig. 24 its model being shown in

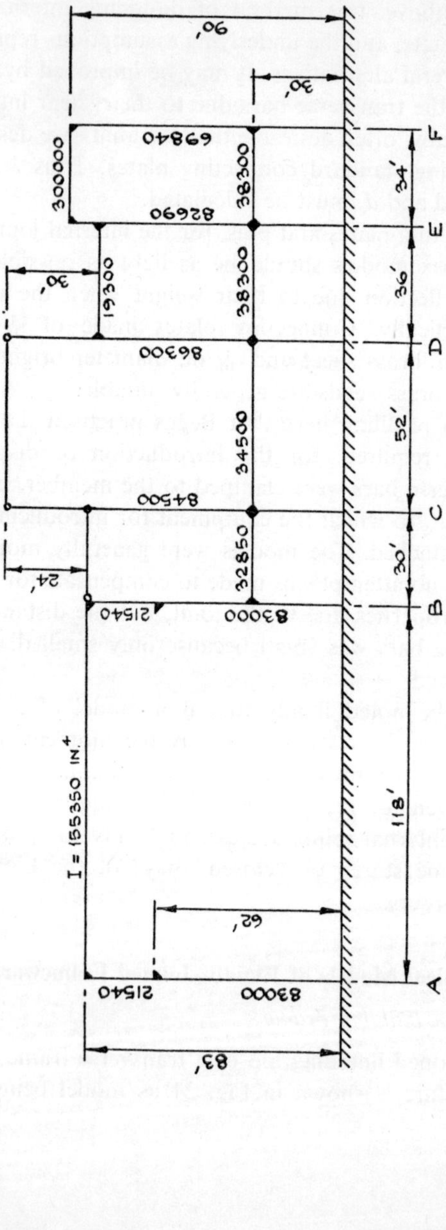

90'
30'
300000
69840
82690
38300
34'
F
E
19300
30'
38300
36'
86300
D
34500
52'
84500
C
24'
21540
32850
30'
83000
B
I = 155350 IN.4
118'
62'
21540
83000
A
83'

FIG. 24

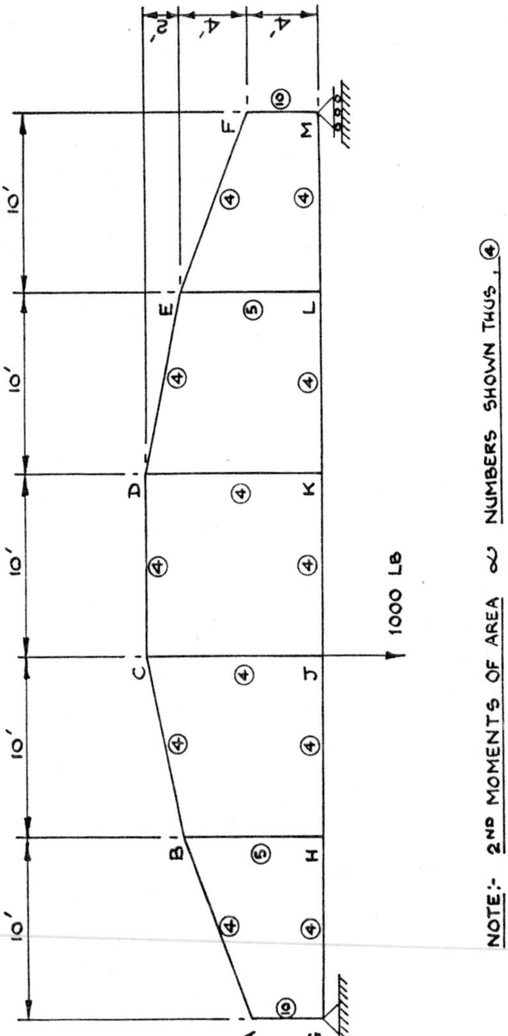

A B C D E F
G H J K L M
10' 10' 10' 10' 10'
2' 4' 4'
1000 LB
NOTE:- 2ND MOMENTS OF AREA ∝ NUMBERS SHOWN THUS, ④

FIG. 25

Fig. 16. The model is made of $\frac{1}{8}$ in. thick perspex sheet to a layout scale of 1 to 50 and a cross-sectional dimension scale of 1 to 100, i.e. $s = 50$: $r = 100$. The joints within members for the purpose of introducing internal displacements have the dimensions $\lambda = 0{\cdot}75$ in., $l_c = 1{\cdot}75$ in. and $d_c = 0{\cdot}19$ in. (by eqn. 4.6). Brass plates $\frac{1}{16}$ in. thick and $\frac{1}{16}$ in. diameter steel pins complete the joints. The method of mounting the model on a vertical baseboard is by means of plates attached to sliders in a T groove along the lower edge of the board, as described above. The model would have been more satisfactory for the purpose of introducing internal displacements if it had been made of sheet celluloid, $\frac{3}{16}$ in. thick to a rather larger layout scale.

Vierendeel Girder

A model of the Vierendeel girder shown diagrammatically in Fig. 25 with relative values of second moments of area of members marked and encircled was made and tested with moment indicators and direct loading in the Civil Engineering Laboratories of the Queen's University of Belfast by students.* The model was cut from $\frac{3}{16}$ in. thick Xylonite celluloid sheet to a layout scale of one-twelfth, i.e. $s = 12$, so that each panel was 10 in. long and a depth of 0·25 in. was chosen for the most slender members having a relative second moment of area of 4 in order to provide the greatest flexibility practicable for the chosen layout scale. Thus the depth of the members of heaviest section, relative second moment of area was 10, was 0·34 in.

The model as mounted vertically for testing is shown in Fig. 26. It was held in position by two pins driven into the baseboard, one at each end of the span. One pin was located in a hole in the model while the other was located in a slot in order to preserve freedom for overall shortening due to flexure of the main members.

Some of the results of the model analysis are shown in Fig. 40 (Chapter 6) and compared with those of moment-distribution calculations for the same condition of loading.

* T. G. Harrison and G. Hull, 1964.

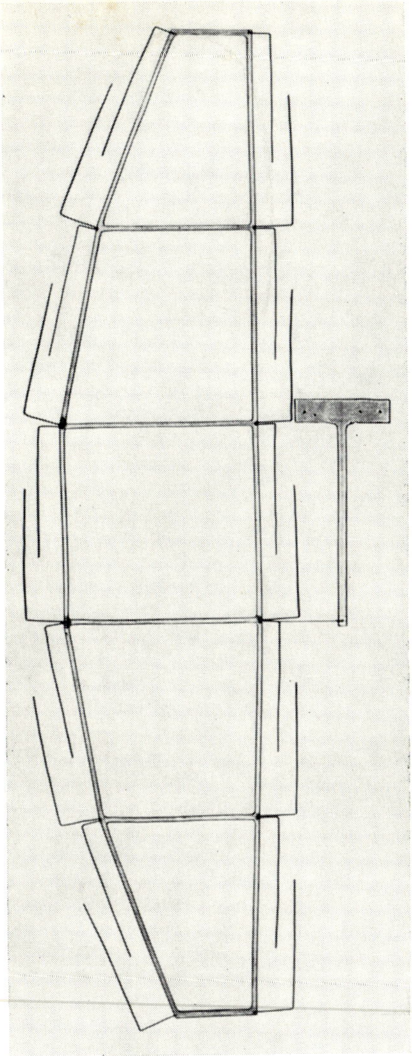

FIG. 26

(*Facing p.* 58)

CHAPTER 5

USE OF MODELS FOR INDIRECT ANALYSIS

In order to describe the testing of a model for the purpose of obtaining force influence lines for its prototype, it is convenient to consider the single-bay portal frame with rigid joints and

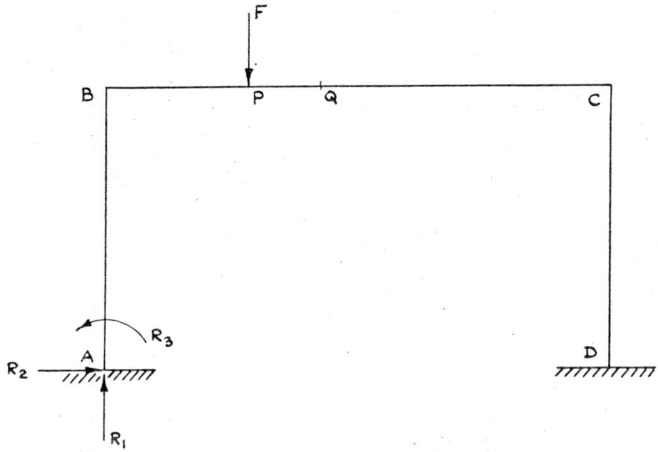

Fig. 27

encastré feet, shown in Fig. 27. The model must be set up in such a manner that the conditions at the feet of the prototype are reproduced ideally. It is essential to take account of the layout scale factor s of the model, but the scale factor for the cross sections of the members or their second moments of area does not

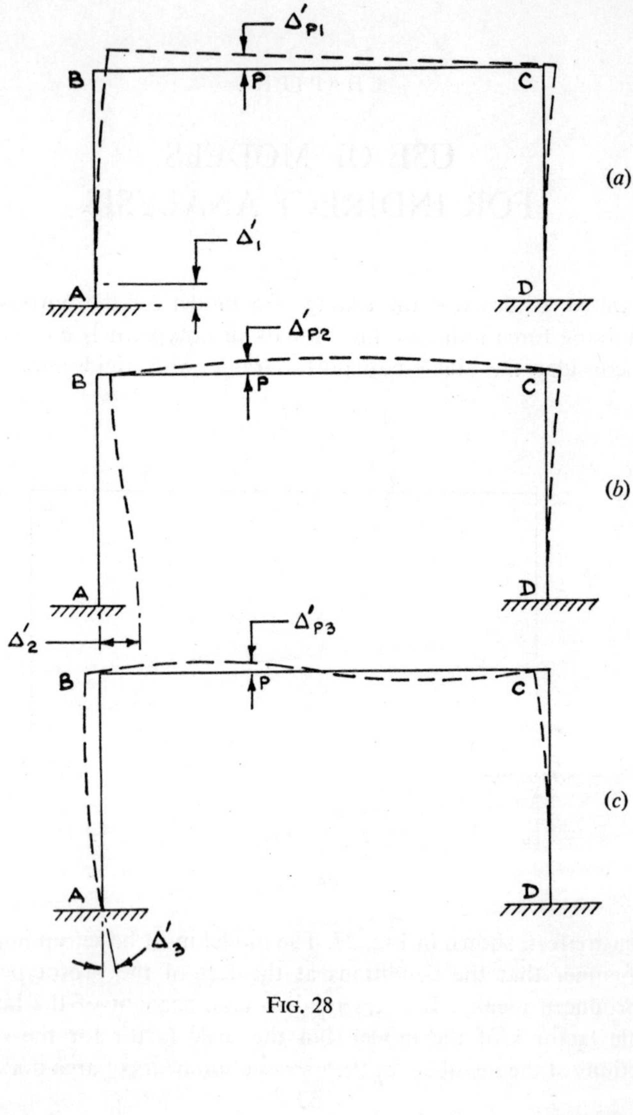

FIG. 28

appear in the analysis since deformation due to bending only of members is significant.

Influence Lines for Reactions

If, say, the influence line for the reaction component R_1 on the prototype at A is required, a small displacement Δ_1' is introduced in model at A as shown in Fig. 28(*a*), care being taken that no other displacements occur at the same time either at A or B. In accordance with Müller–Breslau's principle and the principles of similarity, the small change of configuration of the model due to the displacement represents the required influence line to scale. In order to find the reaction R_1 of the prototype due to a force F applied at P as shown in Fig. 27 the line of action of the force is drawn to scale on the influence line, and the ordinate Δ_{P1}' is measured along the force line as shown in Fig. 28(*a*). Since the ordinates of the influence line would have been s times greater had it been obtained by introducing the scale displacement $s\Delta_1'$ to the prototype

$$R_1 s\Delta_1' - Fs\Delta_{P1}' = 0$$

or

$$R_1 = \frac{s\Delta_{P1}'}{s\Delta_1'} F = \frac{\Delta_{P1}'}{\Delta_1'} F, \tag{5.1}$$

by the principle of virtual work or the Maxwell–Betti theorem (Chapter 3). The influence values for the vertical reaction at A of the prototype are thus simply obtained by multiplying the ordinates of the influence line obtained with the model by $1/\Delta_1'$.

Similarly, the horizontal reaction R_2 on the prototype at A due to the force F is given by:

$$R_2 = \frac{\Delta_{P2}'}{\Delta_2'} F, \tag{5.2}$$

where Δ_{P2}' is the appropriate ordinate of the influence line obtained by introducing a small displacement Δ_2' to the model as shown in Fig. 28(*b*).

Note. In Figs. 28, 29 and 31 the deformed configurations of the model are exaggerated. See Fig. 32 for measurement of ordinates when displacements are large.

The influence line for the moment reaction at A, R_3 is obtained by recording the deformed shape of the model when an angular displacement Δ'_3 is introduced at A, as shown in Fig. 28(c). Since rotation is a dimensionless quantity, if the rotation Δ'_3 were introduced at the foot A of the prototype, by virtue of flexural similarity, deformation curve or influence line with ordinates s times greater then those of the model curve would be obtained. By the principle of virtual work, therefore, the moment R_3 due to the force F is given by

$$R_3 = \frac{s\Delta'_{P3}}{\Delta'_3}, \tag{5.3}$$

where Δ'_{P3} is the ordinate of the model curve measured in the direction of F at the scale position of its point of application. The ordinates of the influence line obtained by the model must, then, be multiplied by s/Δ'_3 to obtain the required influence values for the prototype. Furthermore, if the ordinates Δ'_{P3} are measured in inches, and Δ'_3 is in radians, the moment influence values will be in lb-in. or ton-in. depending upon the unit of force employed.

In the event of the load at a point P of the structure being a couple, instead of measuring an influence-line ordinate it is, of course, necessary to measure the slope of the deflection curve which represents the relevant influence line. Suppose then that Δ'_{P1} is the change of slope at P of the model due to a displacement Δ'_1, then due to a couple M applied at P of the prototype $R_1 = \Delta'_{P1}M/s\Delta'_1$. The scale factor s is necessary now because a change of slope $\Delta_{P1} = \Delta'_{P1}$ of the prototype would be caused by a displacement of A of the prototype $\Delta_1 = s\Delta'_1$.

Internal Forces and Moments

In order to obtain the influence lines for the internal forces and moments at a section Q of the horizontal member of the portal frame shown in Fig. 27 it is necessary to introduce the appropriate internal displacements at Q of the model by providing a joint at this section (see Fig. 22).

The determination of influence lines for the axial and shear forces at Q is similar to the determination of the influence lines for horizontal and vertical reactions. If an axial displacement Δ'_{Q1}

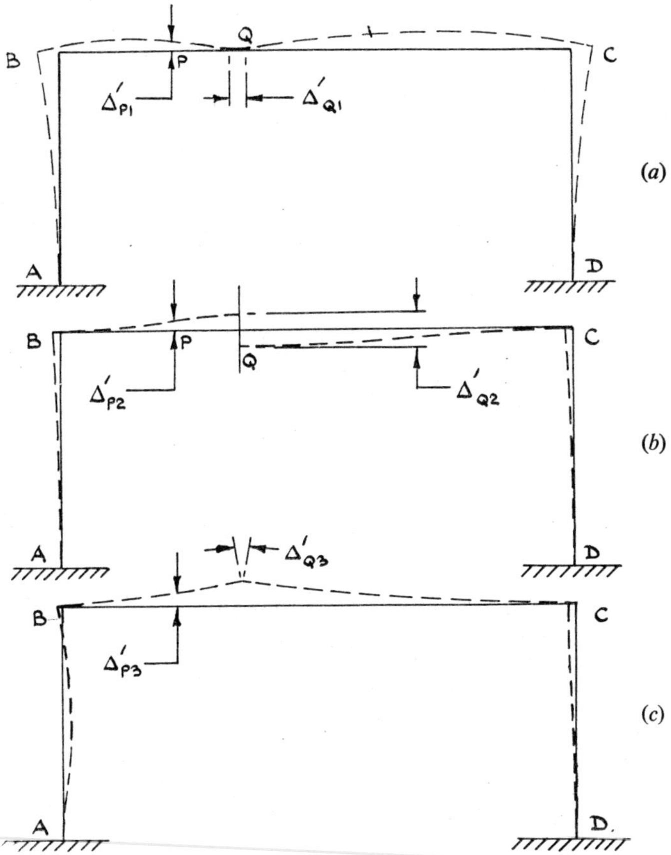

FIG. 29

is introduced at the joint, as shown in Fig. 29(a), in such a way that no shear or rotational displacements occur, then the resulting deformed shape of the model represents the influence line for the axial force T_Q. Similarly, the influence line for the shear force S_Q is obtained by introducing a shear displacement Δ'_{Q2} as shown in Fig. 29(*b*). The magnitudes of T_Q and S_Q when the force F is applied at P are given by:

$$T_Q = \frac{\Delta'_{P1}}{\Delta'_{Q1}} F, \tag{5.4}$$

and

$$S_Q = \frac{\Delta'_{P2}}{\Delta'_{Q2}} F, \tag{5.5}$$

respectively, where Δ'_{P1} and Δ'_{P2} are the respective influence line ordinates at the scale position of P, in the direction of F.

The influence line for the bending moment M_Q at Q is obtained by introducing a change of slope or relative rotation at the joint, as shown in Fig. 29(*c*), and recording the deformed shape of the model. The relative rotation Δ'_{Q3} must not, of course, be accompanied by either axial or shear displacements when this operation is performed. The bending moment M_Q due to the force F applied to the prototype at point P is given by:

$$M_Q = \frac{s\Delta'_{P3}}{\Delta'_{Q3}}, \tag{5.6}$$

where Δ'_{P3} is the appropriate ordinate of the influence line obtained from the model. The derivation of this relationship is exactly the same as that of eqn. (5.3).

Treatment of Distributed Loads

The use of the influence lines for determining the effects of distributed loads is worthy of special attention. If a distributed load of intensity w is applied to the prototype between the points X_1 and X_2, say, as shown in Fig. 30 the virtual work equation for, say, the vertical reaction R_1 at A is:

$$R_1 s\Delta'_1 = \int_{x_1}^{x_2} w(s\Delta'_{X1})\,\mathrm{d}x, \tag{5.7}$$

where Δ_1' and Δ_{X1}' are respectively the vertical displacement introduced into the model at A and an ordinate of the influence line so obtained. Thus:

$$R_1 = \frac{1}{\Delta_1'}\int_{x_1}^{x_2} w\Delta_{X1}' \, dx. \tag{5.8}$$

It should be noted that eqn. (5.8) is merely the integral of eqn.

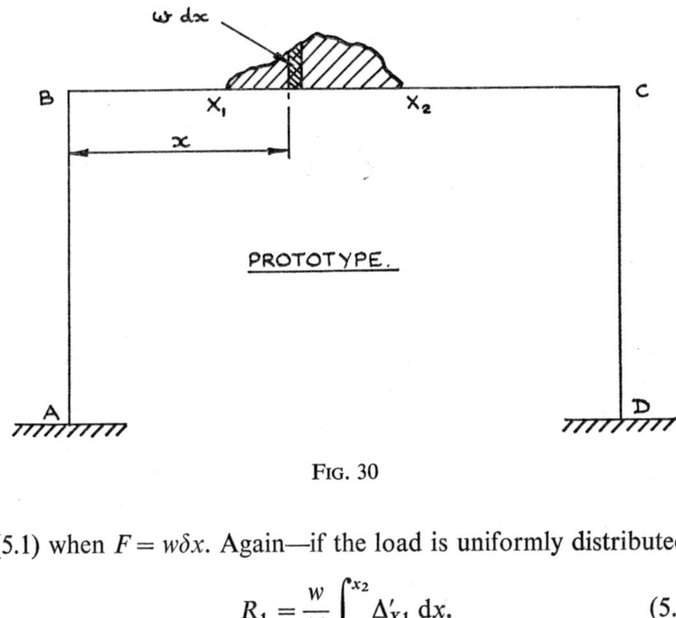

FIG. 30

(5.1) when $F = w\delta x$. Again—if the load is uniformly distributed:

$$R_1 = \frac{w}{\Delta_1'}\int_{x_1}^{x_2} \Delta_{X1}' \, dx. \tag{5.9}$$

Now the integral of eqn. (5.9) can be evaluated by measuring the area enclosed by the influence line provided by the model, between the scale positions of X_1 and X_2 as shown in Fig. 31. The area must, however, be multiplied by the layout scale factor s in order to convert the baseline measurements into the corresponding quantities x of the prototype. That is, if the area obtained from the model is α_1' then $R_1 = sw\alpha_1'/\Delta_1'$.

Similarly:

$$R_2 = \frac{1}{\Delta_2'} \int_{x_1}^{x_2} w\Delta_{X2}' \, dx, \tag{5.10}$$

$$R_3 = \frac{s}{\Delta_3'} \int_{x_1}^{x_2} w\Delta_{X3}' \, dx. \tag{5.11}$$

Similar equations for the other reactions and the internal forces and moment at any section of a member may be obtained likewise.

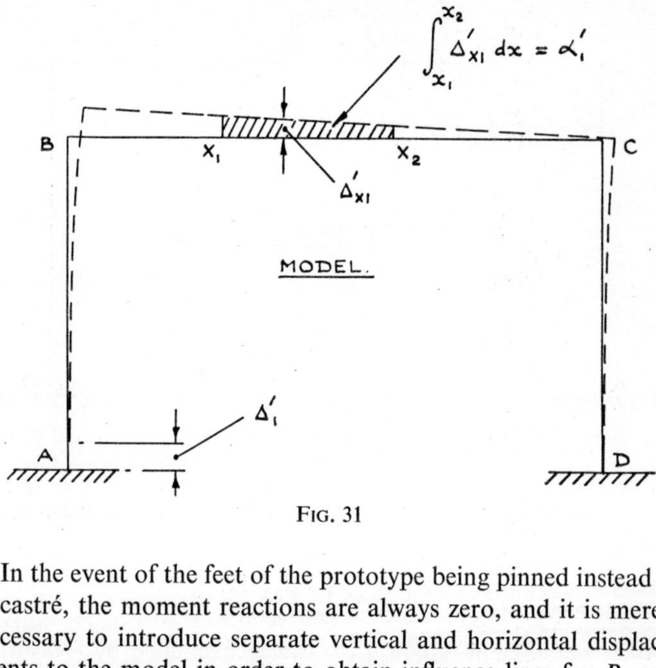

FIG. 31

In the event of the feet of the prototype being pinned instead of encastré, the moment reactions are always zero, and it is merely necessary to introduce separate vertical and horizontal displacements to the model in order to obtain influence lines for R_1 and R_2 respectively.

The Large Displacement Technique

Though the use of large displacements of at least 0·5 in. facilitates the recording of the deformed shapes of models (which represent influence lines for their respective prototypes) on paper by means of a pencil without visual aids, such displacements usually cause significant geometrical distortion of the model and consequent errors in the shapes of the influence lines obtained. These errors can, however, be eliminated for practical purposes by introducing each displacement in both positive and negative senses and measuring the ordinates of the resulting deformation curves in such a way that the component due to geometrical

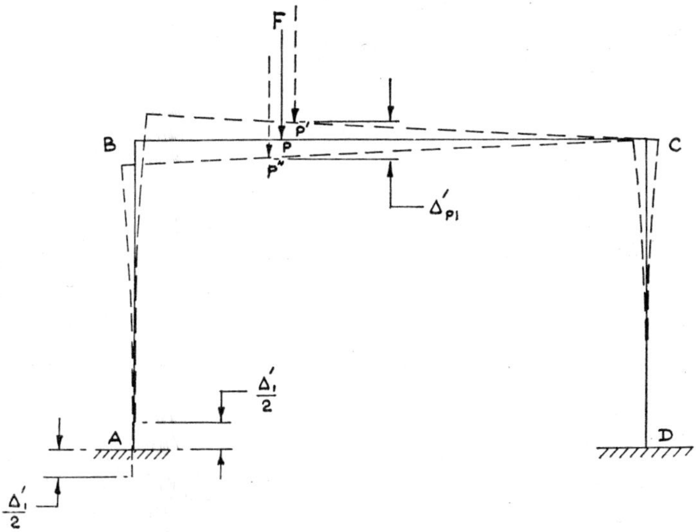

FIG. 32

distortion is diminished. The procedure is best described with the aid of a simple example. The model of a single-bay rigidly jointed portal frame with encastré stanchion feet is shown in Fig. 32 where P is any point on the structure. When the upward vertical displacement $\Delta'_1/2$ is introduced at A the point P assumes the position P' in space while when the displacement is introduced in the opposite sense it assumes the position P''. The appropriate influence line ordinate at P is now obtained by drawing lines parallel with the line of action of a load F at P on the prototype through P' and P'', and then measuring the distance between P' and the projection P'' upon the line through P'. The same result is, of course, obtained if the distance between P'' and the projection of P' upon the line though P'' is measured. If the line of action of F is as shown in Fig. 32 the vertical reaction at the foot A of the prototype due to a load F at P is, then:

$$R_1 = \frac{\Delta'_{P1}}{\Delta'_1} F \tag{5.12}$$

A further feature of the method is that the introduction of equal displacements in both senses enables magnification of the ordinates of the influence lines by nearly two times, to be obtained for a specified maximum permissible displacement introduced in either sense.

When large displacements are introduced into a model, particular care should be taken to ensure that the specified fixing conditions do not change at any point during the recording of the deformation curves. This can be done by marking at a few points as soon as a displacement has been introduced and checking that there has been no change before restoring the model to normal. It is particularly important to mark models in this way at their supports, if the fixing is by clamping screws as described in Chapter 4.

Serious errors due to, say, overstraining or defects in the model material, may be detected by repeating the whole of a test introducing the displacements in a different sequence.

Checking of Model Results

While it is easy to verify, if desired, that the model analysis of a simple structure gives results which are within a few per cent of the values determined by rigorous mathematical analysis founded upon the same basic assumptions, if such a check were attempted for complicated structures the time required for the calculations would be prohibitive and the purpose of the model test defeated. A good indication of the reliability of the results of a model test may, however, be obtained by verifying that the conditions of equilibrium for the structure as a whole and for its individual components are satisfied (see Appendix III).

Examples of Calculations Relating to Model Tests

EXAMPLE 1: Positive and negative horizontal displacements of 0·5 in. of abutment A of the model of the two-pin elastic arch shown in Fig. 33 cause the point Q at the crown of the model arch to have a total vertical component of movement of 0·4 in.

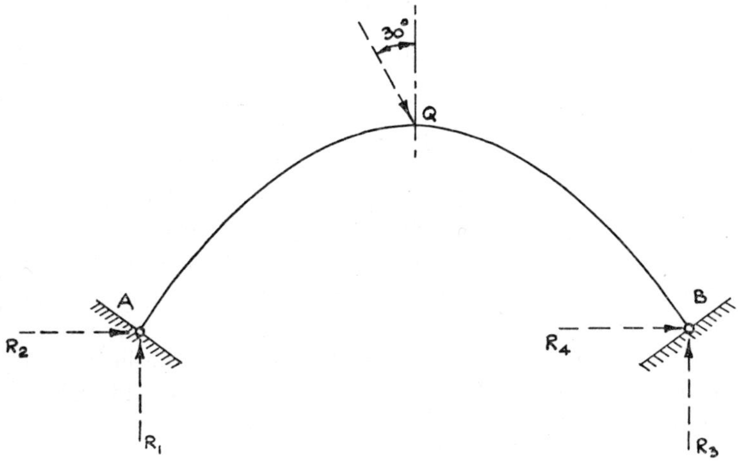

FIG. 33

What horizontal force would the abutments of any geometrically similar full-scale arch sustain due to a vertical concentrated load of 10 ton at the crown of the arch? If the horizontal component of the total movement of Q of the model due to a total horizontal displacement of 1 in. at A, is 0·6 in. and the full scale arch carries a concentrated load of 10 ton at the crown acting in a line 30° to the left of vertical, as shown, what are the horizontal components of the reactions at the abutments?

(*a*) By Müller–Breslau's principle and the principle of virtual work:

$$R_2 \times s \times 2 \times 0{\cdot}5 = 10 \times s \times 0{\cdot}4$$

where R_2 is the full-scale abutment thrust and s is the layout scale factor of the model. Therefore:

$$R_2 = -R_4 = \underline{4 \text{ ton}}$$

(*b*) $R_2 \times s \times 2 \times 0{\cdot}5 + 10 \sin 30 \times s \times 0{\cdot}6 = 10 \cos 30 \times s \times 0{\cdot}4$

Therefore $R_2 = 3{\cdot}464 - 3{\cdot}0 = \underline{0{\cdot}464 \text{ ton}}$

Also, at B by the condition for equilibrium:

$$\begin{aligned} R_4 = 10 \sin 30 + R_2 &= 5 + 0{\cdot}464 \\ &= \underline{5{\cdot}464 \text{ ton}} \end{aligned}$$

EXAMPLE 2: A model of a portal framework with linear elasticity to a scale of 1:30 is shown diagrammatically in Fig. 34. Points 7 in., 14 in., 21 in., and 28 in. from D on stanchion CD move sideways by total distances of 0·38 in., 0·52 in., 0·70 in. and 0·90 in. respectively, due to positive and negative rotational displacements of 0·1 rad introduced at D. Estimate the bending moment at D of the prototype due to (*a*) a concentrated load of 2 ton applied horizontally at C, acting from left to right and (*b*) a horizontal uniformly distributed (wind) load of 0·2 ton ft over the whole length of CD, acting left to right.

(*a*) Due to positive and negative rotational displacements of 0·1 rad of D of the prototype, the total horizontal movement of C would be $30 \times 0{\cdot}9 = 27$ in. since the layout scale factor s of the model is 30. By Müller–Breslau's principle and the principle of virtual work, therefore, for a horizontal load of 2 ton at C, the bending couple at D is:

$$R_6 = \frac{2 \times 27}{2 \times 0{\cdot}1} = \underline{270 \text{ ton-in.}} \text{ (anti-clockwise)}$$

(*b*) Due to positive and negative rotational displacements of 0·1 rad of D of the prototype, the total horizontal movements of points 7 × 30 in., 14 × 30 in. and 28 × 30 in. above the stanchion foot D would be 0·38 × 30 in., 0·52 × 30 in., 0·70 × 30 in. and 0·90 × 30 in., respectively. Now, if y is the height above D of any

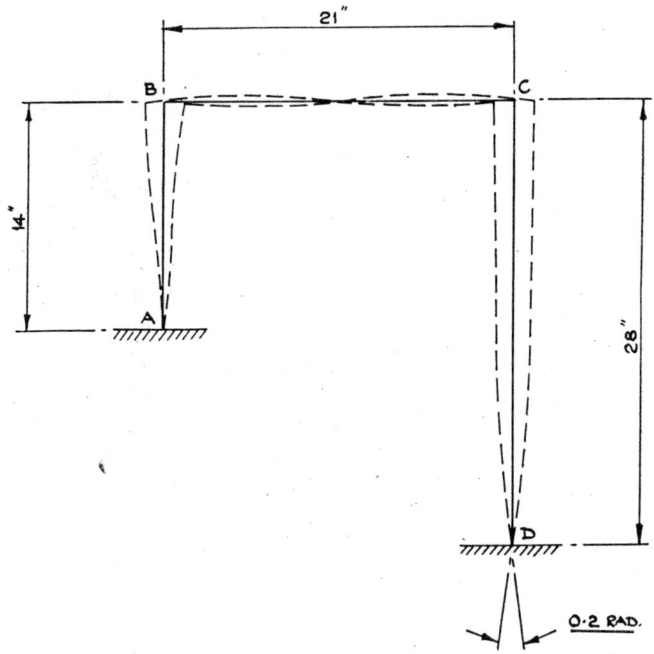

FIG. 34

point on CD of the prototype where the total horizontal movement due to the specified rotational displacement of D would be Δ_y, by Müller–Breslau's principle and the principle of virtual work:

$$R_6 = \frac{w}{2 \times 0{\cdot}1} \int_0^l \Delta_y \, dy = \frac{0{\cdot}2}{2 \times 0{\cdot}1 \times 12} \alpha \text{ ton-in.},$$

where l is the length of CD in inches, w is the intensity of the distributed load in ton/in. and α is the area which would be "swept" by CD during the course of its total horizontal movement. Assuming linear variation of movement between the four points on CD for which the total movement is specified:

$$\alpha = [0{\cdot}38 + (0{\cdot}52 - 0{\cdot}38) + (0{\cdot}70 - 0{\cdot}52) + (0{\cdot}90 - 07{\cdot}0)]$$

$$\times \tfrac{7}{2} \times 30^2 + [0{\cdot}38 + 0{\cdot}52 + 0{\cdot}70] \times 7 \times 30^2 \text{ in}^2$$

$$= 14{\cdot}35 \times 30^2 \text{ in}^2$$

Therefore $R_6 = \dfrac{0{\cdot}2 \times 14{\cdot}35 \times 30^2}{0{\cdot}2 \times 12} = \underline{1076 \text{ ton-in.}}$ (anti-clockwise).

PROBLEMS

1. The member PR of the upper boom of a scale model of the framework shown in Fig. 4 is shortened by $\frac{1}{2}$ in. Calculate the compatible downward displacement of the joint O, hence calculate the force in the member due to a downward force of 2 ton acting at O and state whether it is tensile or compressive. The panels of the framework are square.

Ans.: $\frac{1}{2}$ in.: 2 ton compressive.

2. When the rotational displacement of D of the model described in Example 2 is as shown in Fig. 34, total vertical displacements of 0·25 in. and 0·40 in. occur at points of the beam 3 in. and 6 in. from B, respectively. Estimate the magnitude and sense of the reactive couple at D of the prototype due to a uniformly distributed downward load of 0·5 ton/ft between B and a point 15 ft from B on the beam BC.

Ans.: 253 ton-in. (anti-clockwise).

CHAPTER 6

USE OF MODELS FOR DIRECT ANALYSIS

PRECISELY the same type of model can be used for direct analysis of linear plane frames as are used for obtaining influence lines for forces as described in Chapter 5. Thus for portal frames, models made of celluloid or perspex with different scale factors for the cross-sections of members and layout respectively, can be used, while for trusses metal models with flexible elements in the members are suitable. The models can be mounted vertically to eliminate friction effects as described in Chapter 4, in fact the same mounting equipment can be used for both the direct and indirect testing.

Loading of Models—The Compensating Balance

Unless celluloid or bakelite models (including metal models with celluloid flexible elements) are loaded by the application of a fixed deflection by means of a compensating balance, sufficient time must be allowed for creep to cease. For metal models, however, dead loads can be applied by spring balances or weights suspended by steel wires. (Where pulleys are required to guide the load wires, it might not be sufficient to assume that the weight values represent the true loading.)

The compensating balance for models subject to creep enables the difficulties and inconvenience of dead loading systems as well as those of applying the appropriate value of Young's modulus for the calculation of forces or stresses in the prototype, to be eliminated. It consists of a flexible member made of the same material as the model and of form such that its load–deflection characteristics can be calculated easily. The model is loaded by connecting it to the balance by, say, metal wire, the length of which is such as to cause a fixed total displacement to be shared between the model and balance. Thus the application of dead load is eliminated and the deflection of the balance together with deflections measured on the model provide the necessary information. It is shown below that the deflections of the balance and model remain constant regardless of creep by virtue of the nature of the creep of celluloid. The principle of the compensating balance is, however, useful even when models are made of materials which do not creep.

A convenient form of compensating balance is a cantilever cut from the same sheet of material as the model. Such a balance is illustrated in Fig. 26. Thus the elastic and creep properties of the balance and model are identical. It is essential that the load line is rigid in comparison with the model and balance; otherwise the creep compensation may be upset. Steel wire is suitable for this purpose, and its attachment to a model can be achieved by piercing the model as required, taking care to choose a point or points at which a small hole will not lead to fracture or affect the behaviour of the model significantly.

For complex loading, it is often preferable to apply each component load to the model separately, and use the principle of superposition for finding the effect of the combined system of loading. Again, the relevant influence values for concentrated loads at each joint may be determined by the model to enable the effect of any system of loading to be obtained. Compensating balances can, of course, clearly be used in large numbers to reproduce complex loading. Then each of the concentrated loads is applied through a separate compensating balance, and the deflections of the balances adjusted until they bear the correct ratios to one another. Thus, if all of the balances are identical, their deflections must be in the ratios of the respective prototype loads. Uniformly distributed loads can be reproduced approximately by a balance connected to several points in the manner shown in Fig. 35, the deflection of the balance being then proportional to the total load applied.

A compensating balance should be sufficiently flexible to give a working deflection which is sufficiently large to be measured accurately by an ordinary scale graduated to $\frac{1}{100}$ in. The flexibilities of the model and the balance must, therefore, be of the same order of magnitude; as the flexibilities of the former are

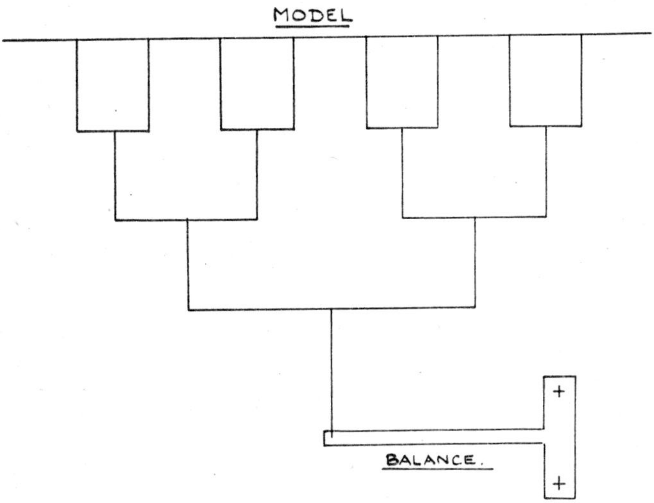

FIG. 35

unknown, the proportions of the latter must be settled by trial in an arbitrary manner.

Suppose that a fixed deflection Δ_f is introduced (Fig. 36) within the load line such that the deflections of the model and balance are Δ' and Δ_b respectively ($\Delta'_f = \Delta' + \Delta'_b$). Then, by virtue of the creep properties described in Chapter 4:

$$\Delta' = K' \frac{F'}{E'}, \tag{6.1}$$

and since the balance is made of the same material as the model:

$$\Delta_b' = K_b \frac{F'}{E'}, \tag{6.2}$$

where K' and K_b' are constant and F' the force in the load line

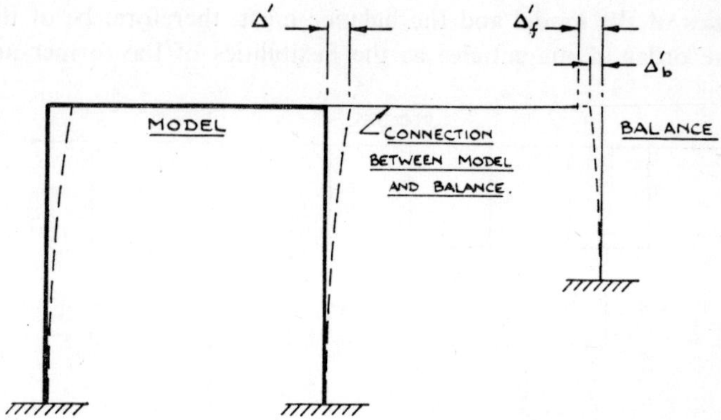

FIG. 36

and E' are variable because of creep of the model and balance. Since:

$$\Delta_f' = \Delta' + \Delta_b', \tag{6.3}$$

then

$$\frac{F'}{E'} = \frac{\Delta' + \Delta_b'}{K' + K_b'} = \frac{\Delta_f'}{K' + K_b'}. \tag{6.4}$$

Thus, for a given total deflection Δ_f' the quantity F'/E' is constant so that Δ' and Δ_b' are constant also by eqns. (6.1) and (6.2).

For a compensating balance of the cantilever type:

$$K_b' = \frac{l_b'^3}{3I_b'} \tag{6.5}$$

where l_b' is the length of the cantilever and I_b' its relevant second moment of area.

Interpretation of Compensating Balance Readings—The Concept of the "Prototype Balance"

Imagine that the load is applied to the prototype through a balance by the introduction of a fixed deflection Δ_f within the load line. If the prototype is made of, say, steel, the imaginary balance must be considered to be made of steel also and flexurally similar to the model compensating balance, so that the ratio of its deflection Δ_b under load to the deflection Δ of the prototype structure due to the same load is equal to the ratio Δ'_b/Δ' of the respective model quantities.

Now, having regard to the properties of constructional materials which obey Hooke's law a fixed load would, for practical purposes, be required to maintain the deflections Δ and Δ_b. In order that the prototype strains shall be similar to those of the model, the magnitude of this load must be such that the deflections Δ and Δ_b of the prototype and balance are equal respectively to $s\Delta'$ and $s\Delta'_b$, i.e. the scaled-up values of Δ' and Δ'_b respectively. If the prototype balance has a constant K_b and a stiffness defined by $b_b = E/K_b$ (where E is the Young's modulus of the prototype and its imaginary balance) the required load can be found immediately as follows:

$$F = b_b\Delta_b = b_b s\Delta'_b. \tag{6.6}$$

Thus the load on the prototype which produces the strain obtained from the model test can be found by eqn. (6.6) when the constant K_b of the imaginary compensating balance of the prototype is known.

The constant K_b of the prototype balance can be found from K'_b by the conditions for flexural similarity. If the same scale factor s is used for model throughout, the prototype balance must be s times larger than the model balance, and it is possible to express K_b as K'_b/s because $K'_b \propto l'^3/I'$. Again, if the model layout scale factor is s, while the scale factor for the cross-sections of the members is r (second moment of area scale factor r^4), the prototype balance must be s times longer than the model balance, but its relevant second moment of area must be r^4 times greater than that of the model balance. In this case $K_b = s^3 K_b'/r^4$. The stiffness b_b of the prototype balance can, therefore, be expressed as:

$$b_b = \frac{sE}{K_b'}, \tag{6.7}$$

if $r = s$; or again as

$$b_b = \frac{r^4 E}{s^3 K_b'}. \tag{6.8}$$

The prototype force is, therefore, given by:

$$F = b_b \Delta_b = \frac{s^2 E}{K_b'} \times \Delta_b', \tag{6.9}$$

or

$$F = \frac{r^4 E}{s^2 K_b'} \times \Delta_b', \tag{6.10}$$

depending upon whether one or two scale factors are involved.

Examples of the calculation of prototype loads from model compensating balance readings are given in Examples 1 and 2 below.

Influence Lines for Deflection

By means of Mohr's theorem based on the reciprocal theorem, as described in Chapter 3, influence lines for deflection of frameworks may be obtained by scale model tests. For example, the deflected configuration of the Vierendeel girder show in Fig. 25, due to a load at joint J, with reference to the undeflected configuration, represents the influence line for the vertical deflection of J. To obtain the influence line it is merely necessary to set-up a scale model of the girder and apply a load at J by means of a compensating balance as shown in Fig. 26. The resulting curve of deformation may be recorded by pencil and paper and the ordinates converted to full-scale values by multiplication by the scale factor of the model. (Magnification of the ordinates may, of course, be obtained by applying the load first in one sense and then in the opposite sense.) Thereafter it is necessary to calculate the prototype value of the load applied to the model at J from the compensating balance deflection, in the manner described above. This is because the deflection influence values or coefficients for J are represented by the deflection curve of the prototype due to a unit load applied at J, in accordance with principles set out in Chapter 3.

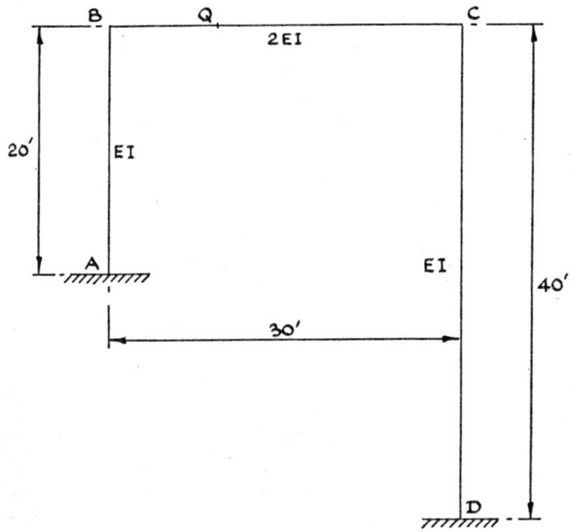

FIG. 37

Calculations relating to influence lines for deflection obtained by model test are illustrated by the following examples.

EXAMPLE 1: The rigidly jointed portal framework shown in Fig. 37 has columns AB and CD of flexural rigidity $EI = 10^7$ ton in^2 and a beam BC of flexural rigidity $2EI = 2 \times 10^7$ ton-in^2. A celluloid model of the framework is loaded at C in the line of BC by a cantilever compensating balance of length 12 in. and flexural rigidity one-half that of the columns AB and CD, of the model. The layout scale factor of the model, $s = 20$, and the member cross-section scale factor, $r = 50$. Determine the sway deflection of C of the prototype due to a concentrated vertical load of 10 ton applied at a point Q on BC, if the vertical deflection of the corresponding point of the model is 0·1 in. when the deflection of the compensating balance is 0·6 in. (Note that the deformation curve of the model due to the loading at C in the line BC represents, to scale, the influence line for side-sway deflection of C of the prototype.)

Length of prototype balance: $l_b = 12 \times 20$ in. $= 240$ in.

Flexural rigidity of prototype balance: $EI_b = 5 \times 10^6$ ton-in^2.

Deflection of prototype balance if applying load F_C to the prototype corresponding to deflection of 0·6 in. of the model compensating balance: $\Delta_b = 0{\cdot}6 \times 20$ in. $= 12$ in.

Deflection at Q of the prototype due to F_C: $\Delta_Q = 0{\cdot}1 \times 20$ in. $= 2$ in.

Now, for the prototype balance:

$$b = \frac{F_C l_b^3}{3\left(\frac{EI}{2}\right)}$$

Therefore

$$12 = \frac{240^3 F_C}{3\left(\frac{EI}{2}\right)}$$

$$F_C = \frac{3 \times 12EI}{2 \times 240^3} \text{ tons, if } EI \text{ is in ton in}^2.$$

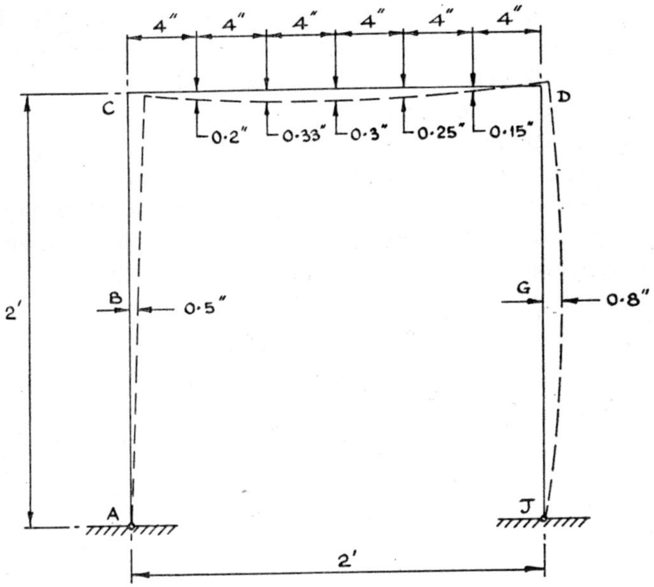

FIG. 38

$$\text{Hence: } \frac{\Delta_Q}{F_C} = a_{QC} = a_{QC} = \frac{2 \times 2 \times 240^3}{3 \times 12EI},$$

so that, due to the 10 ton load at Q, the deflection at C of the prototype would be:

$$\Delta_C = 10a_{CQ} = 10 \times \frac{240^3}{9EI} = \frac{10 \times 240^3 \times 10^{-7}}{9},$$

Therefore $\Delta_C = 1{\cdot}54$ in.

NOTE: When, as in this example, the relative flexural rigidities of the balance and members of the framework can be used, the cross-section scale factor r is irrelevant.

EXAMPLE 2. The deflected form of a celluloid model of a rigidly jointed steel framework with pinned column feet, due to a concentrated side or horizontal load applied by a cantilever type compensating balance at the point G is shown in Fig. 38*. The layout scale factor, $s = 50$ and the member cross-section scale factor, $r = 100$. The compensating balance cut from 0·18 in. thick celluloid sheet is 0·6 in. wide and has an effective length of 15 in. Its tip deflection when applying the load causing deformation of the model as shown is 0·5 in. Determine (*a*) the equivalent loading of the prototype and (*b*) the change in the span BG of the prototype due to a concentrated side load of 10 ton applied at G and a uniformly distributed load of 0·2 ton/ft over the beam CD respectively.

(*a*) For the prototype balance:

$$\text{length } l_b = 50 \times 15 \text{ in.} = 750 \text{ in.}$$

$$\text{second moment of area } I_b = \frac{0{\cdot}18 \times 0{\cdot}6^3}{12} \times 100^4 \text{ in}^4.$$

$$\text{deflection } \Delta_b = 0{\cdot}5 \times 50 = 25 \text{ in.}$$

Now:

$$\Delta_b = \frac{Fl_b^3}{3EI_b} \qquad \left(\Delta_b = K_b \frac{F}{E}\right)$$

Therefore

$$F = \frac{3 \times 25 \times 0{\cdot}18 \times 0{\cdot}6^3 \times 100^4 E}{750^3 \times 12}$$

where E is the Young's modulus of the prototype and is 13,000 ton/in^2. for steel, so that the load applied to the prototype at G

$$F = 750 \text{ ton}$$

(*b*) Due to the 10 ton side load:

$$\Delta_{BG} = 50(0{\cdot}8 - 0{\cdot}5) \times \frac{1}{750} \times 10 \text{ in.} = 0{\cdot}20 \text{ in.},$$

since on the basis of the results of the model test $F = 750$ ton at G would cause a change in the span BG of the prototype of 50 (0·8 – 0·5) in. where $50 = s$, the layout factor of the model.

*Note that the deflections of CD shown are the *true* vertical displacements of points at 4 in. intervals (see Fig. 32).

Due to the distributed load of $w = 0{\cdot}2$ ton/ft over the beam CD:

$$\Delta_{BG} = 2\int_C^D a_{GX} w \, dx$$

since the framework is symmetrical. Now the flexibility coefficient a_{GX} for the deflection of G due to unit vertical load any point X of CD may be obtained from the relationship $a_{GX} = a_{XG}$ on the basis of the reciprocal theorem. Thus if ΔX is the vertical deflection of the prototype at X on CD, due to a horizontal force F at G:

$$a_{XG} = a_{GX} = \frac{\Delta_X}{F}.$$

Substituting above:

$$\Delta_{BG} = \frac{2w}{F}\int_C^D \Delta_X \, dx = \frac{2w}{F} s^2\alpha',$$

where α' is the area enclosed by the deflected and undeflected lines of CD of the model and may be found with sufficient accuracy by assuming the ordinates given to be jointed by straight lines. In this way $\alpha' = 4{\cdot}48$ in^2., so that:

$$\Delta_{BG} = \frac{2 \times 0{\cdot}2}{12} \times \frac{50^2 \times 4{\cdot}48}{750} = 0{\cdot}50 \text{ in.}$$

Therefore, due to both conditions of loading acting simultaneously

$$\Delta_{BG} = 0{\cdot}20 + 0{\cdot}50 = 0{\cdot}70 \text{ in.}$$

The Moment Indicator for Portal Frames

The moment indicator developed at M.I.T. by A. C. Ruge and E. O. Schmidt is a very simple piece of apparatus as shown in Fig. 39. It enables bending moments in the members of models of the structures for which bending only is significant, to be related directly to loading. The instrument consists of two L shaped arms, or pointers, which are attached to a uniform member at points A and B, as shown in such a way that their longer legs of equal length overlap by one-third of the distance AB and are initially parallel to one another and to the member. The indicator arms may be made of metal wire and attached to the member either by the provision of holes into which the wire is a tight fit as shown in Fig. 26, or, alternatively, by clamps.

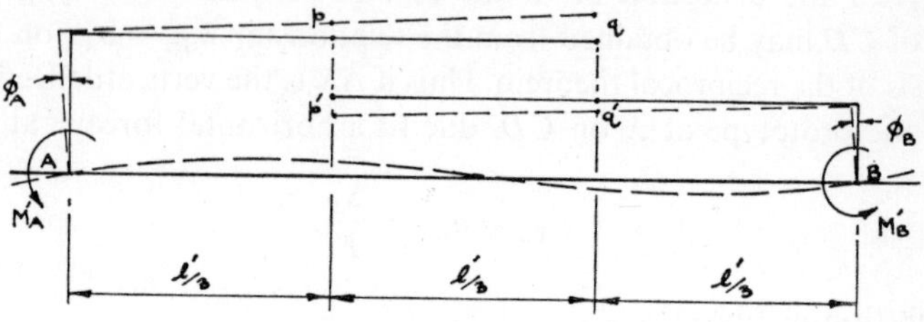

FIG. 39

When the model is loaded, and the member to which the indicator is attached bends, the distances between the points p and p' and q and q' change by amounts depending on the rotations ϕ_A and ϕ_B at A and B, respectively due to bending. If these changes are $\Delta'_{pp'}$ and $\Delta'_{qq'}$ respectively, then by eqn. (3.44):

$$\Delta'_{pp'} = \frac{l'}{3}\phi_A + \frac{2l'}{3}\phi_B = \frac{l'}{3}(\phi_A + 2\phi_B),$$
$$\Delta'_{qq'} = \frac{2l'}{3}\phi_A + \frac{l'}{3}\phi_B = \frac{l'}{3}(2\phi_A + \phi_B), \qquad (6.11)$$

provided that the angles ϕ_A and ϕ_B are small and l' is the distance between the feet of the indicator arms. The angles ϕ_A and ϕ_B are related to the bending moments M'_A and M'_B at A and B respectively, the points of attachment of the indicator arms, as follows (Chapter 3):

$$M'_A = \frac{2E'I'}{l'}(2\phi_A + \phi_B),$$
$$M'_B = \frac{2E'I'}{l'}(\phi_A + 2\phi_B). \qquad (6.12)$$

Comparison of eqns. (6.11) and (6.12) shows that it is possible to substitute:

$$M'_A = \frac{6E'I'}{l'^2}\Delta'_{qq'},$$
$$M'_B = \frac{6E'I'}{l'^2}\Delta'_{pp'} \qquad (6.13)$$

The proportions of the indicator should be such that the quantities $\Delta'_{pp'}$ and $\Delta'_{qq'}$ can be measured accurately by a scale without optical aids after they have been recorded on paper behind the model.

It should be noted, however, that eqns. (6.12) as given, are applicable to uniform members only provided that there is no load applied between points *A* and *B* (see Chapter 3). For this reason, the moment indicator is particularly useful in circumstances where there are appreciable lengths of member which may be regarded as free from load or again where the actual or equivalent loading (as described in Chapter 3) may be regarded as concentrated loads applied at joints. Vierendeel bridge girders and rigidly jointed building frameworks subjected to wind loading provide examples of circumstances in which the moment indicator is likely to be particularly useful.

Interpretation of Moment Indicator Results for the Prototype

Probably the best way of interpreting the moment-indicator readings obtained by the model test is to convert the deflections $\Delta'_{pp'}$ and $\Delta'_{qq'}$ into the corresponding prototype values by multiplying them by the layout scale factor *s* of the model. The moments at the corresponding points of the prototype member are, therefore, given by:*

* Note that for the indicator as shown in Fig. 39, $\Delta'_{qq'}$ and $\Delta'_{pp'}$ are positive when the distances *qq′* and *pp′* increase, corresponding to anti-clockwise bending couples applied to the ends of the member.

$$M_A = \frac{6EI}{l^2}\, s\Delta'_{qq'},$$

$$M_A = \frac{6EI}{l^2}\, s\Delta'_{pp'}, \tag{6.14}$$

where $l = sl'$ (and $E = 13{,}000$ ton/in^2 approximately if the prototype is made of steel). These equations are appropriate even if the scale factor for the cross-section of the model members is different from the layout scale factor s, because the deflected form of the model is nevertheless similar to that of the prototype for similar loadings.

Having found the bending moments for the prototype in this way, the corresponding loading may be found from the compensating balance readings as described above. Alternatively, they may be found by means of the relative values of the second moments of area of the structural members and the compensating balance, as in the example below.

Some typical results* of analysis by moment indicator are shown in Fig. 40(*a*) for a Vierendeel girder (described in Chapter 4), while the results of analysis by Wessman† by means of moment distribution are shown in Fig. 40(*b*).

EXAMPLE: Moment indicator deflections with respect to the ends of the uniform member *AB*, length 10·77 in., of a one-twelfth scale model of the Vierendeel girder shown in Fig. 25 when loaded vertically at *J* by means of a cantilever type compensating balance were $\Delta'_{pp'} = 0{\cdot}19$ in. and $\Delta'_{qq'} = 0{\cdot}35$ in., respectively, for a compensating balance (tip) deflection of 1·05 in. The model and balance were cut from 0·18 in. thick sheet celluloid, the depth of the model member *AB* being 0·25 in. while the length and depth of the balance were 10 in. and 0·44 in., respectively. Determine the values of the bending moments in the member *AB* at *A* and *B*, respectively, of the prototype due to a concentrated vertical load of 1000 lb applied at *J*.

* Obtained by T. G. Harrison and G. Hull. † See Appendix I, Ref. 29.

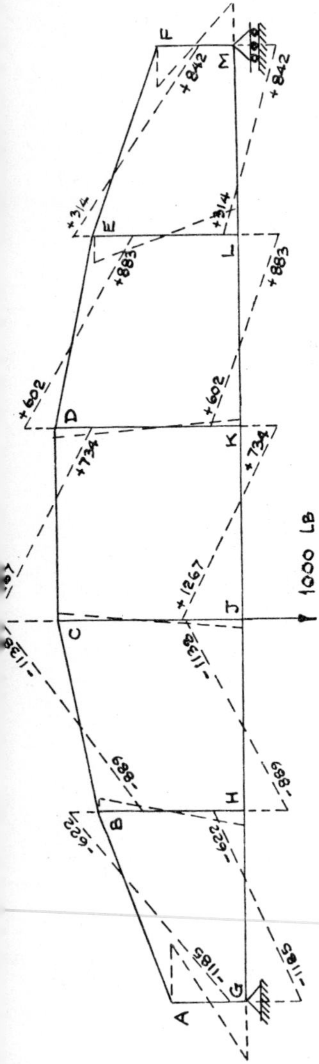

FIG. 40 (*a*)

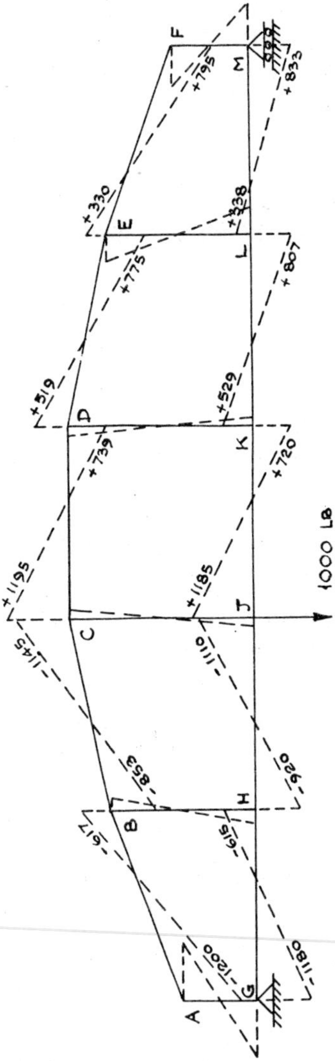

FIG. 40 (*b*)

BENDING MOMENT IN LB. FT.

Consider the prototype balance and let its deflection correspond to scale that of the model balance, i.e. $\Delta_b = 1{\cdot}05 \times 12$ in. Further, suppose that for this deflection the load of 1000 lb is applied at J. Then, since the deflection of the tip of a cantilever due to a load F there, is:

$$\Delta = \frac{Fl^3}{3EI}.$$

For the prototype balance:

$$1{\cdot}05 \times 12 = \frac{1000(10 \times 12)^3}{3EI_b},$$

Therefore $$I_b = \frac{10^6 \times 12^3}{3 \times 1{\cdot}05 \times 12E}.$$

For the member AB, therefore:

$$I = \left(\frac{0{\cdot}25^3}{0{\cdot}44}\right) I_b = \frac{10^6 \times 12}{3 \times 1{\cdot}05E}\left(\frac{0{\cdot}25}{0{\cdot}44}\right)^3$$

since the depth of the model member AB is 0·25 in. while the depth of the compensating balance is 0·44 in.

Now, by eqn. (6.14) for the member AB of the prototype:

$$M_A = \frac{6EI}{l^2}\, s\Delta'_{qq'}$$

$$M_B = \frac{6EI}{l^2}\, s\Delta'_{pp'}.$$

Substituting $l = 12 \times 10{\cdot}77$ in., $\Delta'_{pp'} = 0{\cdot}19$ in., $\Delta'_{qq'} = 0{\cdot}35$ in. and the expression for I derived above:

$$M_A = \frac{6 \times 10^6 \times 12^2}{10{\cdot}77^2 \times 12^2 \times 3 \times 1{\cdot}05}\left(\frac{0{\cdot}25}{0{\cdot}44}\right)^3 \times 12 \times 0{\cdot}35 = 12{,}600 \text{lb-in.}$$

$$M_B = \frac{6 \times 10^6 \times 12^2}{10{\cdot}77^2 \times 12^2 \times 3 \times 1{\cdot}05}\left(\frac{0{\cdot}25}{0{\cdot}44}\right)^3 \times 12 \times 0{\cdot}19 = 6820 \text{ lb-in.}$$

Thus, there is an anti-clockwise couple of 12,600 lb-in. acting on the member at A and an anti-clockwise couple of 6820 lb-in. acting on the member at B.

Alternatively

For the prototype system, since $s = 12$:

$\Delta_b = 12 \times 1{\cdot}05$ in.; $\Delta_{pp'} = 12 \times 0{\cdot}19$ in.,; $\Delta_{qq'} = 12 \times 0{\cdot}35$ in.;

$I_b = (12 \times 0{\cdot}44)^3 t$; $l_b = 12 \times 10$ in.; $I = (12 \times 0{\cdot}25)^3 t$;

$l = 12 \times 10{\cdot}77$ in.

where t is the thickness of the sheet celluloid from which the model and compensating balance were made. Then:

$$\Delta_b = 12 \times 1{\cdot}05 = \frac{F(12 \times 10)^3}{3(12 \times 0{\cdot}44)^3 tE}.$$

Therefore, if $F = 1000$ lb:

$$\Delta_b = \frac{10^6}{3 \times 0{\cdot}44^3 tE},$$

and:

$$\Delta_{pp'} = \frac{0{\cdot}19}{1{\cdot}05} \times \frac{10^6}{3 \times 0{\cdot}44^3 tE},$$

$$\Delta_{qq'} = \frac{0{\cdot}35}{1{\cdot}05} \times \frac{10^6}{3 \times 0{\cdot}44^3 tE}.$$

Hence:

$$M_A = \frac{6(0{\cdot}25 \times 12)^3 tE}{(10{\cdot}77 \times 12)^2} \times \frac{0{\cdot}35}{1{\cdot}05} \times \frac{10^6}{3 \times 0{\cdot}44^3 tE} = \underline{12{,}600 \text{ lb-in.}}$$

$$M_B = \frac{6(0{\cdot}25 \times 12)^3 tE}{(10{\cdot}77 \times 12)^2} \times \frac{0{\cdot}19}{1{\cdot}05} \times \frac{10^6}{3 \times 0{\cdot}44^3 tE} = \underline{6820 \text{ lb-in.}}$$

The Moment Deformeter

The theoretical principle underlying the so-called moment deformeter for obtaining influence lines for curvature (and hence bending moment) of a structure by model test, is described in Chapter 3. The deformeter is simply a means of applying a known system of loads, as shown in Fig. 10(*d*), at the part of the model for which the influence line for curvature is required. It consists of a light, rigid frame (made, say, of aluminium alloy), a balance beam of second moment of area I_b' made of the same material as the model and a circular plug for introducing a concentrated load between the balance beam and the model, as shown in Fig. 41. When the plug is pushed into place there is a central deflection of

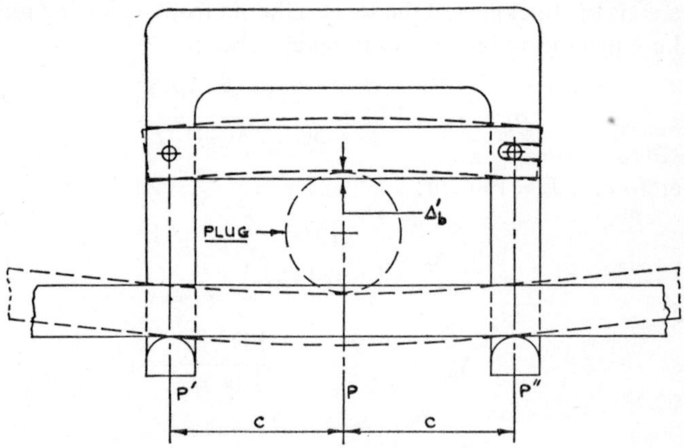

FIG. 41

the balance beam Δ_b' and a deflection of any point of the model outside the region $P'P''$ of application of the deformeter, Δ'. Thus the curvature R_P' at P due to a concentrated load proportional to Δ_b' applied at the point in the direction of Δ' is:

$$\frac{1}{R_P'} = \frac{2\Delta'}{C^2}, \tag{6.15}$$

by eqn. (3.39).

For the corresponding point P of the prototype whose dimensions are s times greater than those of the model, the curvature due to a deflection $\Delta_b = s\Delta_b'$ of a balance beam of the same material of the prototype and in a deformeter s times larger than the deformeter applied to the model, would be:

$$\frac{1}{R_P} = \frac{2s\Delta'}{s^2C^2} = \frac{2\Delta}{s^2C^2}, \tag{6.16}$$

where $\Delta = s\Delta'$. This curvature would be caused, therefore, by a concentrated load at any point acting in the direction of Δ, the deflection of the point caused by the "prototype deformeter":

$$F = \frac{48EI_b}{(2sC)^3}\Delta_b = \frac{6E(s^4I_b')}{s^3C^3}s\Delta_b', \tag{6.17}$$

if the balance beam is simply supported, since $\Delta_b = s\Delta_b'$ and $I_b = s^4I_b'$ (or r^4I_b'). Hence for unit load applied anywhere on the prototype, the curvature at P would be:

$$\frac{1}{R_P} = \frac{2\Delta'}{sC^2}\frac{C^3}{6Es^2I_b'\Delta_b'} = \frac{C}{3s^3EI_b'}\frac{\Delta'}{\Delta_b'}, \tag{6.18}$$

where E is the modulus of elasticity of the material of which the prototype is made. The bending moment at P due to unit load is obtained merely by multiplying $1/R_P$ by EI (where I is the second moment of area of the *uniform* member upon which point P is located).

It should be noted that the moment deformeter is based upon an approximate relationship for the curvature of a finite length of a structural member. It requires further investigation for use as a "large displacement" device.

Coker's Method for Direct Measurement of Model Strains

In a paper read before the Institution of Civil Engineers in 1929 Professor E. G. Coker dealt with a direct method of model analysis which involved the measurement of lateral strains. It is well known that for a state of plane stress the orthogonal principal stresses σ_1 and σ_2 produce a lateral strain given by:

$$\varepsilon = \frac{\sigma_1 + \sigma_2}{E} \nu, \tag{6.19}$$

where ν is Poisson's ratio. If the elastic material is sufficiently thin, the variation of $(\sigma_1 + \sigma_2)$ between the surfaces can be neglected for practical purposes, and the sum at any point may be assumed to be constant throughout the thickness of the body. The principal stress sum can thus be found by measuring the change of thickness δt and writing:

$$\sigma_1 + \sigma_2 = \frac{\delta t}{t} \times \frac{E}{\nu}. \tag{6.20}$$

If δt is measured at a free boundary for example, at the edge of a member, one of the principal stresses, say, σ_2 is zero so that:

$$\sigma_1 = \frac{\delta t}{t} \frac{E}{\nu}. \tag{6.21}$$

The extreme fibre stresses in a loaded model of a plane framework can, therefore, be found, subject to the above assumptions, by measuring the thicknesses of members at a number of points at the edges before and after loading and substituting the values of $\delta t'$ so obtained, in eqn. (6.21). The moments in the members can be found, if necessary, by the simple Bernoulli–Euler beam theory, having separated the stresses due to direct forces from those due to bending.

If the model has been loaded by the application of a fixed deflection, using a compensating balance, the strains and stresses in the prototype for an equivalent deflection, i.e. $\Delta = s\Delta'$ can be found without difficulty. Thus, if a longitudinal *surface* strain ε' of the model, due to bending, has been found by lateral measurements and the relationship:

$$\varepsilon' = \frac{1}{\nu'} \times \frac{\delta t'}{t'}, \tag{6.22}$$

the corresponding strain of the prototype ε is given by:

$$\varepsilon = \varepsilon' \frac{d}{sd'} = \frac{1}{\nu'} \times \frac{\delta t'}{t'} \times \frac{d}{sd'}, \tag{6.28}$$

where d and d' are the depths of the corresponding members of the prototype and model respectively. (The ratio d/d' is not necessarily the same as the linear scale factor s by virtue of the license in model design afforded when strains due to axial forces are negligible.) If the model is an exact scale replica of the prototype, scale deflections will be associated with identical strains in it. Once the strains in the prototype have been found, the stresses can be obtained merely by multiplication by the modulus of elasticity E.

The magnitude of the load on the prototype which produces a deflection $\Delta = s\Delta'$ and the strains ε can be found by the concept of the "prototype balance", as described above. It is, therefore, easy to find the stresses produced by any value of the load.

The practical difficulty of this method of model-testing lies in the accurate measurement of the lateral displacement $\delta t'$. A lateral extensometer is necessary, which must be capable of detecting 10^{-5} in. The lateral displacement of a piece of celluloid $\frac{1}{8}$ in. thick for a single principal stress of 3000 lb/in^2 taking Young's modulus for celluloid as 3,000,000 lb/in^2 and ν as 0·35 is:

$$\delta t = \frac{3000 \times 0{\cdot}35}{3{,}000{,}000 \times 8} = 0{\cdot}00044 \text{ in.},$$

which is relatively large for this thickness of material in view of the ultimate stress of celluloid being approximately 7000 lb/in^2. Professor Coker designed a lateral extensometer to measure 10^{-6} in., but measurement of quantities of this order of magnitude is clearly very sensitive to temperature variations.

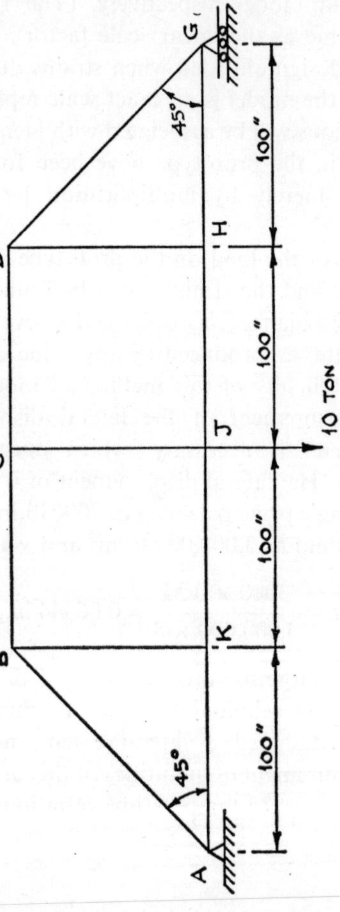

Fig. 42

PROBLEMS

1. Using the data of Example 2 above, calculate the change in the span BG of the portal frame, whose model is shown in Fig. 38, due to a concentrated load of 5 ton acting on the beam CD at a point 33 ft 4 in. from D.

Ans.: $+0{\cdot}186$ in.

2. The members of the rigidly jointed framework shown in Fig. 42 have a flexural rigidity of 10^6 ton-in^2. A celluloid scale model of the framework is subjected to a vertical concentrated load at J by means of a cantilever compensating balance 10 in. long cut from 0·18 in. thick sheet celluloid to a uniform width of 0·44 in. The model is constructed to a scale of 3 to 25. If the deflection of J of the model is 0·47 in. due to a balance deflection of 0·43 in., what is the deflection of J of the prototype due to a load there of 10 ton?

Ans.: 0·24 in.

3. The model of the framework shown in Fig. 42 and described in problem 2 above, is tested using moment indicators. For the indicator arms of member JK attached at J and K deflections $\Delta'_{pp'}$ and $\Delta'_{qq'}$ of 0·38 in. and $-0{\cdot}46$ in., respectively, where the negative sign denotes shortening of the distance between the points of the indicator (p is adjacent to J and q is adjacent to K), correspond to a balance deflection of 0·52 in. The compensating balance is a cantilever 10 in. long and cut from 0·18 in. thick sheet celluloid to a uniform width of 0·44 in. What are the bending moments at J and K of the member JK of the prototype due to a load of 10 ton at J?

Ans.: 11·5 ton-ft (sagging); 9·3 ton-ft (hogging).

APPENDIX I

BIBLIOGRAPHY

Similarity

1. Thomson, J. Comparisons of similar structures as to elasticity, strength and stability. *Collected Papers in Physics and Engineering*, ed. J. Larmor. Cambridge University Press, Cambridge (1912), 361–72.
2. Johansen, F. C. Research in mechanical engineering by small-scale apparatus. *Proc. I. Mech. E.*, **1** (1929), 151–272.
3. Buckingham, E. Model experiments and the forms of empirical equations. *Trans. A.S.M.E.*, **37** (1915), 263–93).
4. Charlton, T. M. A note on the flexural similarity of structures. *Civ. Eng. & P.W. Rev.*, **43** (1948), 456–9.
5. Charlton, T. M. Some recent experimental methods applied to engineering design. *Trans. N.E.C. Inst. Engrs. and Shipbldrs.*, **65** (1948), 29–46.
6. Duncan, W. J. A review of dimensional analysis. *Engineering*, **167** (1949), 533.
7. Langhaar, H. L. *Dimensional Analysis and Theory of Models*, John Wiley, New York (1951).
8. Ipsen, D. C. *Units, Dimensions and Dimensionless Numbers*, McGraw-Hill, New York (1960).

Indirect Methods of Model Analysis

9. Gottschalk, O. Mechanical calculation of elastic systems. *J. Franklin Inst.*, **202** (1926), 61–88.
10. Beggs, G. E. The use of models in the solution of indeterminate structures. *J. Franklin Inst.*, **203** (1927), 375–86.
11. Rieckhof, C. *Experimentalle Statik*. Darmstadt (1927).
12. Gottschalk, O. The experiment in statics. *J. Franklin Inst.*, **207** (1929), 245–60.
13. McCullough, G. B. and Thayer, E. S. *Elastic Arch Bridges*, John Wiley, New York (1931), 269.
14. Beaufoy, L. A. Fundamental difficulties of deformeter analysis. *Engineering*, **134** (1932), 671–3.
15. Lobban, C. H. Mechanical methods of solution of stresses in frames. *Trans. Inst. Engrs. and Shipbldrs.* (*Scotland*), **77** (1933–4), 169–226.

16. PIPPARD, A. J. S. and SPARKES, S. R. The simple experimental solution of certain structural design problems. *Proc. I.C.E.*, **4** (1936–7), 79–92.
17. HAYDEN, A. G. *The Rigid Frame Bridge*, 2nd ed. (Ch. 13 by G. E. Beggs), John Wiley, New York (1940).
18. ENEY, W. J. A large displacement deformeter apparatus for stress analysis with elastic models. *Proc. Soc. Exp. Stress Analysis*, **6**, 2 (1948), 84–93.
19. ENEY, W. J. Studies of continuous bridge trusses with models. *Proc. Soc. Exp. Stress Analysis*, **6**, 2 (1948), 94–105.
20. ROCHA, M. and BORGES, F. Photographic method for model analysis of structures. *Proc. Soc. Exp. Stress Analysis*, **8**, 2 (1951), 129–42.
21. MATHESON, J. A. L. *Hyperstatic Structures*, Vol. 1, Butterworth, London, 1958.
22. CHARLTON, T. M. *Analysis of Statically Indeterminate Frameworks*, (see Ch. 6 and 8), Longmans Green, London (1961).

Direct Methods of Model Analysis

23. COKER, E. G. Some experimental methods and apparatus for determining the stresses in bridges and framed structures. *Proc. I.C.E.*, **229** (1929–30), 33–76. (For details of Coker's lateral extensometer see FILON, L. N. G., *A Manual of Photoelasticity for Engineers*, Cambridge University Press, 1936.)
24. BAKER, J. F. The mechanical and mathematical stress analysis of steel building frames. I.C.E. Selected engineering paper 131 (1932).
25. RUGE, A. C. and SCHMIDT, E. O. Mechanical structural analysis by the moment indicator. *Proc. A.S.C.E.*, **64** (1938), 1613–25.
26. CARDOSO, E. Testing reduced models of structures. *Civ. Eng. & P.W. Rev.*, **42** (1947), 540–1.
27. CHARLTON, T. M. A direct method for the model analysis of structures. *Civ. Eng. & P.W. Rev.*, **48** (1953), 51–3.

General

28. RITTER, M. Experimentelle Methoden der Baustatik. *Schweiz. Bau*, **96** (1930), 226–9.
29. WESSMAN, H. E. Analysis of vierendeel trusses. *Disc. Trans. A.S.C.E.*, **102** (1937), 926–30.
30. NORRIS, C. H. Model analysis of structures. *Proc. Soc. Exp. Stress Analysis*, **1**, 2 (1944), 18–34.
31. ROSS, A. D. Effects of creep on instability and indeterminacy investigated by plastic models. *The Structural Engineer*, **24** (1946), 413–28.
32. SCHMIDT, E. Modellversuche zur Bemessung von Baukonstruktionen. *Schweiz. Bau.*, **67** (1949), 555–61.
33. HETENYI, M. *Handbook of Experimental Stress Analysis*, John Wiley, New York (1950).

34. ROCHA, M. General review of the present status of the experimental method of structural design. Prelim. Publication. Fourth Congress Internat. Assoc. for Bridge and Struct. Eng. (1952), 329–51.

35. SCRUTON, C., WHITBREAD, R. E. and CHARLTON, T. M. An aerodynamic investigation for the 437 ft tower block proposed for the Albert Embankment, Vauxhall, London. *NPL Aero. Rep.* 1032. Nat. Phys. Lab., Teddington (1962).

36. CHARLTON, T. M. *Energy Principles in Applied Statics*, Blackie, London, (1959).

37. NEAL, B. G. *Structural Theorems and Their Applications*, Pergamon, London (1964).

38. HEYMAN, J. *Beams and Framed Structures*, Pergamon, London (1964).

39. PARKES, E. W. *Braced Frameworks*, Pergamon, London (1964).

APPENDIX II

THE PRINCIPLE OF VIRTUAL WORK APPLIED TO FRAMEWORKS

THE application of the principle of virtual work to structures is perhaps best demonstrated by reference to pin-jointed frames, although the principle is of general utility. Thus, the system of forces acting at a point as shown in Fig. 43 might represent the state of affairs at any joint of a loaded pin-jointed frame. If these forces are in equilibrium and the point is given a small arbitrary finite displacement (by an external agency which need not be defined) such that the directions of the forces are unchanged for practical purposes, and equilibrium is maintained, then by the law of conservation of energy the net work done during this process is zero because there is no resultant force acting. That is, if such a displacement of a point in equilibrium under the action of a system of forces is imagined to take place, the virtual work is zero. In mathematical terms this equation of virtual work is as follows:

$$F_1\Delta_1 + F_2\Delta_2 + \ldots + F_N\Delta_N = 0,$$

or

$$\sum^{N} F_j\Delta_j = 0, \qquad (1)$$

where the Δ's are the components of the arbitrary displacement in the lines of action of the forces, respectively. The F's are a system of forces in equilibrium and the Δ's are a system of geometrically compatible displacements. Clearly, those terms in eqn. (1) in which the sense of Δ is opposite to the sense of F will be negative because in such cases work (virtual) is done upon the force.

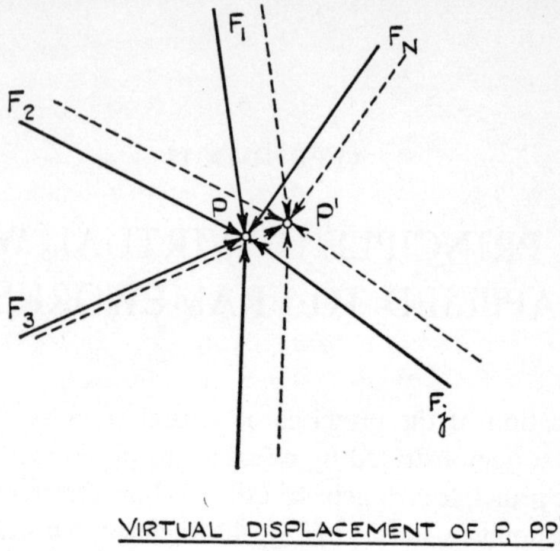

F1
FN
F2
P
P'
F3
Fj
VIRTUAL DISPLACEMENT OF P, PP'

FIG. 43

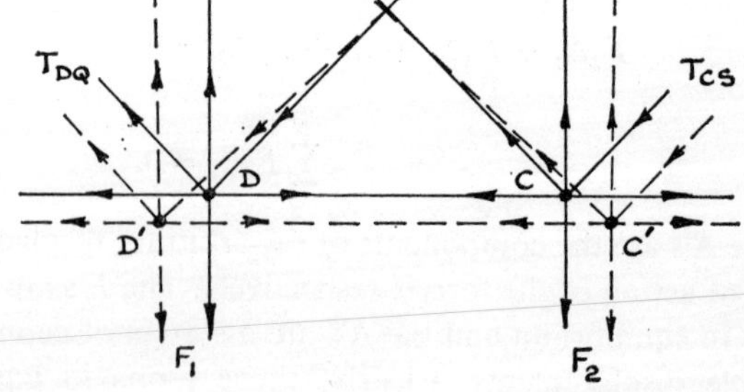

A'
B'
A
B
TAP
TBR
TDQ
TCS
D
C
D'
C'
F1
F2

FIG. 44

Consider now the various joints of a loaded pin-jointed frame as shown in Fig. 44 (a statically indeterminate frame is shown for the sake of generality). At each joint the conditions considered above are represented, namely, a point in equilibrium under the influence of a system of forces. If, then, a small arbitrary displacement is imagined to be given to each joint (e.g. AA', BB', CC', and DD' as shown in Fig. 44), such that the geometry of the loaded structure is not changed significantly (it is emphasized that these imaginary displacements are quite arbitrary and are not necessarily related to those caused by the straining of the members due to the forces applied to them) the following virtual-work equations will be obtained:

(*a*) At joints at which there are no applied loads:

$$\sum T\Delta = 0,$$

(*b*) At joints at which an external load F_j is applied:

$$\sum T\Delta + F_j\Delta_j = 0,$$

where T represents the force in a member and Δ and Δ_j are the components of the virtual displacement of the joint in the line of action of T and F_j respectively. If all the equations of types (*a*) and (*b*) are added, the general equation of virtual work for the whole structure is obtained:

Compatible displacements

$$\sum^{N} T_i\Delta_i = \sum^{N} F_j\Delta_j, \qquad (2)$$

Forces in equilibrium

where Δ_i is now the *total virtual displacement in the line of the i-th member*: that is, the combined effect in that line of the virtual or imaginary displacements of the two joints connected by that member; n is the number of members in the structure and N is the number of external forces or loads applied to the structure.

The equality in eqn. (2) arises from the convention of tension and extension of a member being consistent: that is, a member with tension T_i subjected to a virtual extension Δ_i gives a negative virtual work term $T_i\Delta_i$ since virtual work is done upon the member.

If virtual displacements at supports are considered, then the reactions are included with the member (internal) forces, so that in the most general case n includes the number of reactions as well. It will be noted that the principle is concerned with the elements of statics and involves no assumptions concerning the load–deflections properties of the structure.

It is important to realize that eqn. (2) is valid for any system of forces in the structure which are in equilibrium and any arbitrary set of virtual displacements which are geometrically compatible with the geometry of the loaded structure, i.e. which if actually imposed by, say, adjusting turn-buckles in the members would not cause a change in the geometry of the loaded structure of the first order of magnitude. Care has been taken to refer to the geometry of the "loaded structure"; that is because the structure is being considered as a force system. It is usually a prior requirement that the geometry of a structure after loading shall be substantially the same as that of the unloaded condition; otherwise, large deflections under load are implied. On this basis, for an elastic structure, the changes of length e of the members and the corresponding deflections Δ in the lines of action of the loads due to a system of loading can be chosen as a permissible system of virtual displacements of the loaded system, so that it is possible to write:

Compatible displacements

$$\sum^{n} T_i e_i = \sum^{N} F_j \Delta_j \qquad (3)$$

Forces in equilibrium

where e_i is the result of the force T_i and Δ_j is due to F_j and the other loads. Having regard to the basis of the principle of virtual work it is, however, clearly legitimate to retain the system of compatible displacements due to one system of loads and employ them as the virtual displacements in writing a virtual-work equation for another set of loads F'_j, etc., and resulting member forces T'_i, etc., thus:

Compatible displacements

$$\sum^{n} T'_i e_i = \sum^{N} F'_j \Delta_j \tag{4}$$

Forces in equilibrium.

APPENDIX III

VERIFICATION OF ANALYSIS OF PORTAL FRAMEWORKS

ONE way of checking that analysis of a framework has been performed correctly is to verify that the conditions of equilibrium are satisfied throughout the framework. Perhaps the most convenient way of doing this is by means of the principle of virtual work for which purpose the first step is to determine the number of independent equations of equilibrium which may be expected for a framework whose loading is specified.

Suppose that for a certain framework and specified loading n ordinates of the bending moment diagram are necessary to define it. If, then, the framework has r redundants the number of independent equations of equilibrium relating the n bending moment values and the loading is $(n - r)$. The use of the principle of virtual work for deriving these equations is best described with the aid of simple examples.

First, consider the simple portal framework with encastré feet, loaded as shown in Fig. 45(a) and suppose the couples acting on the members are as shown. The bending moments at A, B, C, D and G are sufficient to define the distribution of bending moment throughout the framework for the particular loading considered, so that $n = 5$. Therefore, since $r = 3$, two independent equations relating bending moments and loads are expected and they may be obtained by considering two independent virtual displacements

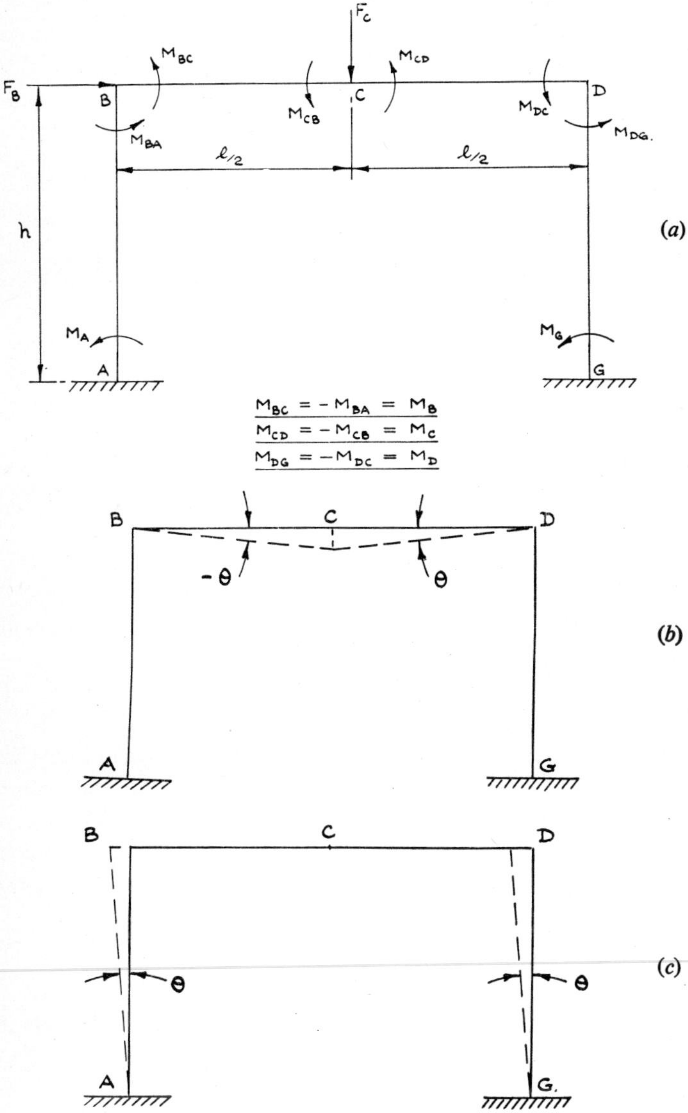

F_C
M_{BC}
M_{CD}
F_B
B
C
D
M_{CB}
M_{DC}
M_{BA}
M_{DG}
$l/2$
$l/2$
h
(a)
M_A
M_G
A
G
$M_{BC} = -M_{BA} = M_B$
$M_{CD} = -M_{CB} = M_C$
$M_{DG} = -M_{DC} = M_D$
B
C
D
$-\theta$
θ
(b)
A
G
B
C
D
(c)
θ
θ
A
G.

FIG. 45

of the loaded framework, as shown in Figs. 45(*b*) and 45(*c*). Then, by virtual work, having regard to Fig. 45(*b*):

$$-M_B\theta + M_C\theta + M_C\theta - M_D\theta + F_C\frac{l}{2}\theta = 0,$$

or (1)

$$-M_B + 2M_C - M_D + F_C\frac{l}{2} = 0,$$

and having regard to Fig. 45(*c*):

$$M_A\theta - M_B\theta + M_D\theta + M_G\theta - F_B h\theta = 0.$$

or (2)

$$M_A - M_B + M_D + M_C - F_B h = 0.$$

Equations (1) and (2) are two independent equations of equilibrium which must be satisfied by the results of the analysis of the framework for the loading shown in Fig. 45(*a*).

Next, consider the single-storey, two-bay portal framework, with encastré feet, loaded as shown in Fig. 46(*a*). The bending moments at *A*, *B*, *C*, *D* (3 values), *G*, *J*, *K* and *L* are sufficient to define the distribution of bending moment throughout the framework for the specified loading, that is $n = 10$. Therefore, since $r = 6$, four independent equations of equilibrium may be expected corresponding, say, to the four independent virtual displacements of the loaded framework shown in Figs. 46(*b*), (*c*), (*d*), and (*e*) respectively.

It will be noted that the virtual displacement shown in Fig. 45(*e*) consists of rotation of the joint *D*. The virtual work of the loading and internal couples due to the virtual displacements shown, considered in turn, gives the following four independent equations of equilibrium (after cancelling the θ's):

$$-M_B + 2M_C + M_{DC} + F_C\frac{l_1}{2} = 0, \quad (3)$$

$$-M_{DG} + 2M_G - M_J + F_G\frac{l_2}{2} = 0 \quad (4)$$

$$M_A - M_B + M_{DL} + M_L + M_J + M_K - F_B h = 0, \quad (5)$$

$$M_{DC} + M_{DL} + M_{DG} = 0. \quad (6)$$

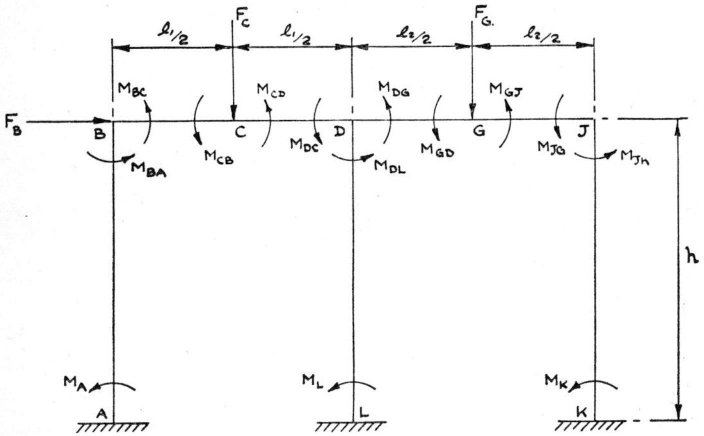

$$M_{BC} = -M_{BA} = M_B \qquad M_{GJ} = -M_{GD} = M_G$$
$$M_{CD} = -M_{CB} = M_C \qquad M_{JK} = -M_{JG} = M_J$$

FIG. 46 (*a*)

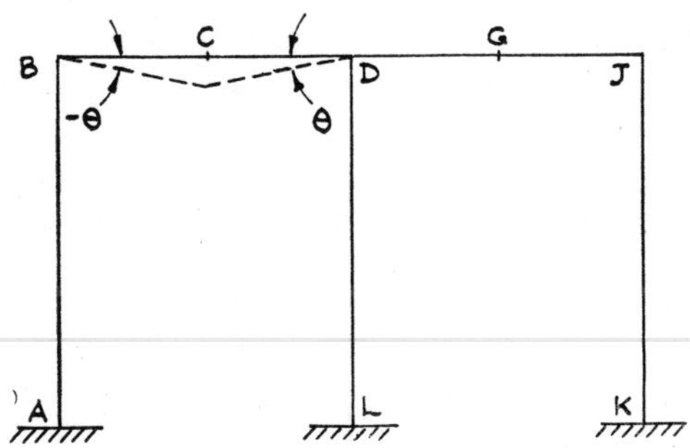

FIG. 46 (*b*)

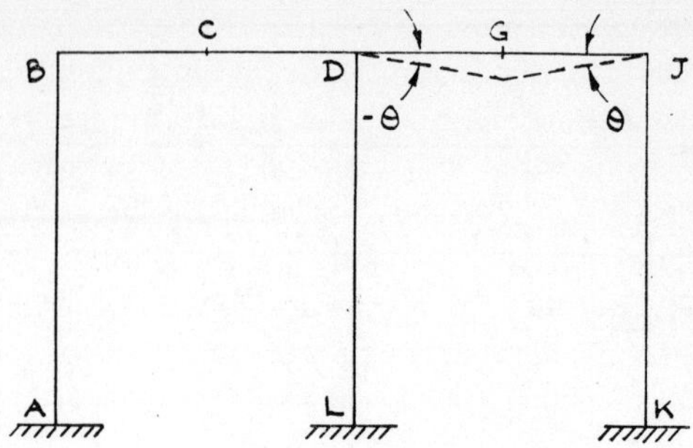

FIG. 46 (*c*)

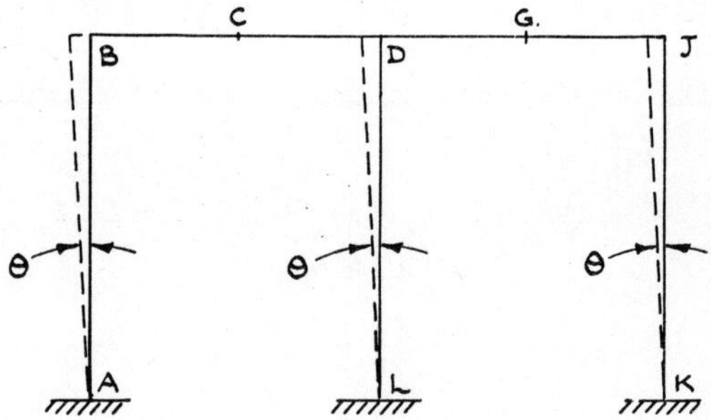

FIG. 46 (*d*)

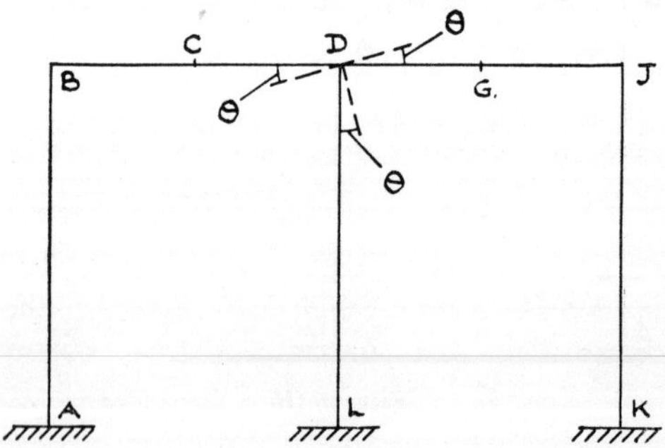

FIG. 46 (*e*)

Again, these equations must be satisfied by the results of analysis of the framework for the loading specified in Fig. 46(*a*).

As a third example, consider the rigidly jointed framework with encastré feet loaded as shown in Fig. 47(*a*). The bending moments at A, B, C, D (3 values), J (3 values), K, L, N, P (3 values), Q, R and S are sufficient to define the distribution of bending moment throughout the framework for the specified loading. Thus, since $n = 18$ and $r = 9$, nine independent equations of equilibrium may be expected, which can be derived by considering the nine independent virtual displacements shown in Figs. 47(*b*) to (*f*) inclusive, and are as follows:

$$M_B - 2M_C + M_{DC} + \frac{w_1 l_1^2}{4} = 0, \tag{7}$$

$$-M_{JQ} - 2M_Q + M_{PQ} + \frac{w_2 l_2^2}{4} = 0, \tag{8}$$

$$M_K - 2M_L - M_N + \frac{w_3 l_2^2}{4} = 0, \tag{9}$$

$$M_A + M_B + M_{DS} + M_S - (M_{JD} + M_{DJ}) \times \frac{h_1}{h_2 - h_1} = F_B h_1, \tag{10}$$

$$(M_{JD} + M_{DJ}) \frac{h_2}{h_2 - h_1} + M_{PR} + M_R = (F_J + F_K) h_2, \tag{11}$$

$$M_{JK} + M_K + M_N + M_{PN} = F_K h_3, \tag{12}$$

$$M_{DC} + M_{DS} + M_{DJ} = 0, \tag{13}$$

$$M_{JK} + M_{JD} + M_{JQ} = 0, \tag{14}$$

$$M_{PN} + M_{PQ} + M_{PR} = 0. \tag{15}$$

By substituting the bending moment values obtained by analysis in these equations the accuracy of the analysis for the specified loading may be verified.

Although any *rigorous* mathematical procedure of structural analysis makes use of equations of equilibrium of the kind considered, either explicitly or implicitly, this does not preclude their

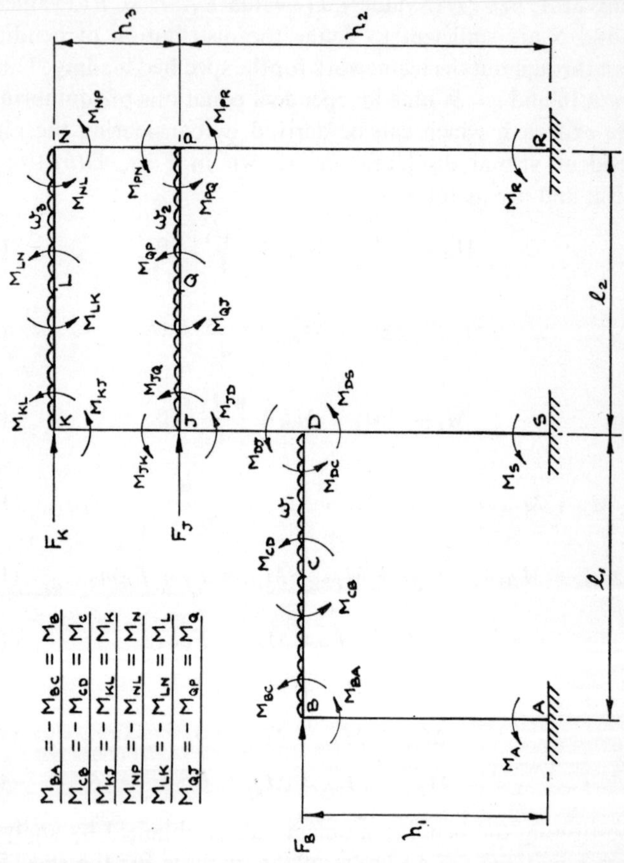

FIG. 47 (*a*)

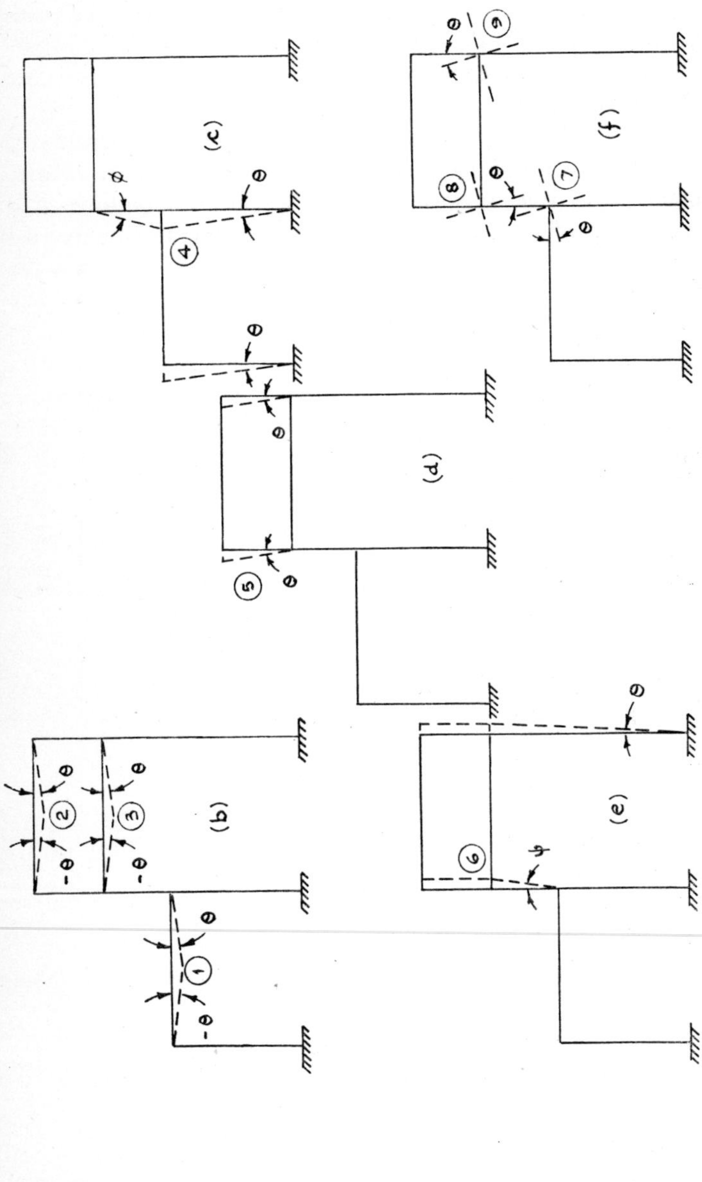

FIG. 47 (*b*) – (*f*)

ultimate use for checking the results of analysis. Thus, the results of analysis might be incorrect due to errors of principle or arithmetical errors or both. The proposed means of checking is particularly useful when either model analysis or approximate analysis is used.

Use of the principle of virtual work is possible regardless of the nature of the framework, that is, the joints can be pinned, semi-rigid or rigid because attention is focused upon forces and couples regardless of their origin. It is interesting to note that the systems of virtual displacements which are used to obtain the independent equations of equilibrium of a particular framework loaded in a specified manner, as described above, correspond to the *actual* primary modes of simple plastic collapse of the framework for the kind of loading considered. The estimation of collapse loads ignoring instability then follows the checking process without difficulty.